human futures

D0862733

human futures

needs · societies · technologies

Special World Conference on Futures Research, Rome, 1973.

FUTURES

special publication
in co-operation with IRADES
(Istituto Ricerche Applicate Documentazione e Studi)

CB158
S65
1973a

Published by Futures,
the journal of forecasting and planning,
IPC Science and Technology Press Ltd.,
IPC House, 32 High Street, Guildford, Surrey, U.K.

ISBN 0 902852 34 5

Typeset by Mid-County Press Ltd., Grosvenor House, 18 The Ridgeway
Wimbledon, London SW19, U.K.

Printed in England by Whitstable Litho, Millstrood Road, Whitstable, Kent.

Contents

Foreword

The Rome Special World Conference on Futures Research held in September 1973 followed three world futures meetings in Oslo (1967), Kyoto (1970) and Bucharest (1972), as a response to an increasing awareness among futurists that the discussion and research on alternatives for the world should be focussed on the principal issue — *human needs.*

The Istituto Ricerche Applicate Documentazione e Studi (IRADES) in Rome was responsible for arranging the Conference. In co-operation with the Continuing Committee for World Future Research Conferences, they defined the theme of the Conference as a link between changing needs, in their various manifestations, according to time, culture and environment, and *new societies* that answer or seek to answer them. *Supportive technologies* to fulfil these needs and ensure the survival of these societies formed the third point of the discussion.

Two problems were identified early in the planning of the Conference. First of all, the theme, by its nature, is boundless and difficult to investigate; hence the Conference required a special organisational design. Secondly, if the investigation is to be meaningful, its conclusions should be translatable into practical applications.

IRADES decided on a self-creating conference. They did not simply draw up a programme to which the participants were asked to adhere: instead the theme was allowed to develop gradually in accordance with the wishes of the majority of those taking part.

At the outset five persons of different countries, cultures and disciplines were asked to suggest the framework. On the basis of their replies, the organisers prepared a questionnaire and submitted it to 20 experts in all parts of the world who, in turn, replied with suggestions for a draft programme. This was then sent to 40 persons for further comments and proposals, and ultimately to 60 others. From this final response, a new programme was prepared and expanded as far as possible to embrace a spectrum of opinions representing different nations, cultures and ideologies, but all united in their concern for the future of man and his needs.

Six studies in this publication are revised versions of papers commissioned by IRADES for the Conference to examine the approaches to, and resolutions of the complex issue of needs, societies and technologies.

These studies by *Maurice Guernier, Lewis Mumford, Sam Cole* and *Craig Sinclair, Jim Dator,* and *William Simon* were reviewed by *John McHale,* whose contribution opened the proceedings of the Conference. *Bertrand de Jouvenel's* "A Word to Futurists" formed part of his introductory speech.

The second aspect of the Conference, the development of the discussions towards a programme for action, is examined by *Harold A. Linstone,* who devised the mechanism of interaction between working groups, which produced their selection of priorities for future action.

Finally, the prospects for futures studies are assessed by *Yehezkel Dror.* His and Dr Linstone's paper were written after the Conference for this collection.

The account of authorship would not be complete without mention of *Pietro Pace* and *Eleonora Masini,* who with others from IRADES prepared this discussion about the content of human futures and ways to them.

A word to futurists

Bertrand de Jouvenel

At the beginning of our century Henry Adams saw the development of human power as the great law of History, inevitable and terrifying. Man disposed and would dispose of increasing energy, and do so at a successively accelerated rate: a mathematician of 1950, he said, would be able to plot the future graph of these forces.

The new American, in his view, would be the product of these forces, and every American destined to live until the year 2000 (like most of you) would know how to control unlimited power. He would think in terms of complexities inconceivable to a mind of the preceding age. He would tackle problems quite outside the scope of the previous society. And Adams imagined this man coming and pondering two centuries after Gibbon (1764) was exploring the meaning of the Roman empire. And it is now nine years after the date he set for it.

The reference to Gibbon is significant. His reflection rested on a known past and tried to explain the causes of it. Ours has a bearing on an unknown future, and tries to discover what we can make it.

This resolution is very different from prediction of a probable future: it is research into possible futures and their nature. These two resolutions proceed from two distinct states of mind, one responding to a demand, the other to a moral impulse, and they require different methods.

The practice of forecasting is natural and useful: it is based simply on the recognition of strong tendencies whose future courses can be anticipated – already being done using naive extrapolation – and then by allowing for interactions between variables taken account of, thus forming a system.

These predictive devices have answered a demand which has increased rapidly, reflecting our heightened perception of change, and has made it desirable to have Baedeker guides to the future. Indeed, everyone with a particular interest, whether narrow or wide, wants to know more about the future environment in which this interest will be acted out.

These predictive devices have met with a certain success. But this success is in the form of essentially conservative visions, based on past changes that, it is believed, must continue, and on an impact of certain variables which have been the centre of attention for quite a long time.

It is a very striking contrast that the methods which, when applied to economic variables, led to the most optimistic visions, once applied to ecological variables, have produced a vision of tragic pessimism.

The lesson to be learned from this is simple: one only sees what one is looking at. Observation is unconsciously selective, guided by knowledge acquired by the observer. The admiral sees clearly that his naval base is impregnable: he is only looking at the sea. The base will fall immediately, when attacked by land.

Forecasting is vital, but the effort of forecasting can be fatal if what one is looking at makes one neglect what should be looked at.

Forecasts proceeding from specific domains are now being effected by the integration of a growing complexity of contributions from different sources into models. These models will be used for simulations allowing us to look for what we are interested in: that is in knowing how to impart favourable inflections to trends.

However, I must confess an anxiety. These great predictive machines will be costly, of experts' time and of equipment. Who will have the means for this? The authorities with their own interests and preferences, which will then come and balance the variables.

The more important forecasting becomes, the more important it is to consider *what* are the questions about the future that research should be directed towards and thus *who* asks the questions. How is one to avoid the biassing of the answers by the interests and learnings of those who set the questions?

It also becomes more important to consider what use is made of the forecasts: whether it is to guide decision makers or to justify them, and in so far as it is to guide them, whether they use this guidance for the general good or for their own advantage.

In other words there is a political aspect to forecasting which has to be considered. The paramount importance of political forecasting derives from the nature and character — particularly moral character — of the authorities.

Forecasting is particularly useful as an early-warning mechanism, but particularly tiresome when it is offered or received as an Established Future — an Authorised Version of the Future. This unacceptable approach serves what I will call the obscurantist religion of the 'forces of Progress' which sweep us along, *nolens volens*. It is positivist mythology involving worship of a 'self-making', a kind of atheistic providence, immanent to societies and even to the human species, administering goods and evils that must be accepted.

It is this blind belief in the esoteric forces of Progress which has caused them to be greeted hastily in all bearers of change, just as the Aztecs believed they recognised Quetzacoatl in Cortez, who was to be their conqueror. This trust in everything that appeared as an agent of change has made us accept all the destructions of what stood in the way of these agents. And appearing like the chosen instruments of a new Providence, how could we grudge the sacrifices they called for? But what was the promise borne by these agents?

A promise of human power which has been kept. Our civilisation has proceeded to an unimagined collective power, and to the concentration of the authorities which control it, be they political or economic. Decision making has been concentrated and individuals released from it; the currency of power is distributed to them in the form of the car at their door, energy is delivered to the home, purchasing power is theirs, and what is more important the worker and the housewife are relieved of physical effort. All good things which should be a cause for our rejoicing but not for our contentment.

But has this progress created a climate favourable to biological man the individual, each man unlike even the most ingeniously thought-out machine, having unknown potentialities, which can be realised given a favourable environment? This is doubtless what Marx had in mind when he said that agents of material progress were unconsciously preparing the transition from the reign of necessity to that of freedom. But where did this transition take place?

Never have men been so dependent for so many aspects of their lives, and dependent on long-term decisions, taken at a level from which they cannot be seen as people.

That is the fact of a society controlled by different organisations whose forms each have their own logic. Those most active in transforming our lives only know us as buyers; others have appeared on the scene to protect us as wage-earners, and, much later on, as independent workers. Those who know us as members of society are relatively very weak. To cap it all, an expanding bureaucracy can only meet increasing demands in proportion to the pressure exerted by each.

We can see misunderstandings developing between the large organisations and the people — a natural result of their different points of view.

The management of a large organisation has by virtue of its position a systemic perception. No army headquarters has ever been aware of each soldier. But the question is precisely that of knowing if each of us is merely a soldier of progress towards collective power or whether it is the people who constitute the focal point of our civilisation.

We are a civilisation that does not build monuments. Everything we construct is destined to rapid destruction to make way for buildings of superior efficiency. But just supposing our buildings have only utilitarian value, could the case be the same with men? Have they not an ultimate value, and if this does not come through in the quality of the monuments to civilisation perhaps it comes through in what men become by virtue of that trouble taken to provide them with a favourable climate for their personal development.

The imperatives of progress have divorced men from nature, divorced them from labour that they could acknowledge as their own work, and has broken the ties formed between them. Moralists will reproach our contemporaries for their acquisitive natures. But isn't the love of objects a transfer of affections which have been thwarted in other ways by abrupt change?

Happy are those like us who can form new attachments through our community of interests! But in this respect we are a privileged minority. Let us use this privilege to seek ways of ending the isolation.

Let us be cautious about imitating the physical sciences, in our social planning, since their quantitative concepts are often not applicable to men.

In the course of recent years it has been shown that the great counsellor, economic science, is, by the character of its concepts, divorced from nature; the relations with nature, which involve no financial transactions, have remained quite foreign to it. Let us take care that our social sciences are not depersonalised, that is, dehumanised.

The time is favourable for changes in this desirable attitude. Indeed we are seeing the discrediting of simplistic predictions (those I have called 'conservative'). This is happening because they are based on a principle of continuity, which is valid over a long period, but we have, in Péguy's terminology, emerged from a 'période' to enter an 'époque' marked by discontinuity, turbulence, and uncertainty: a time of confusion, but because of this, favourable to orientation in new directions. In looking for the best new directions we must recognise that possible futures lie hidden in the present. But account must be taken of ways of bringing these about. You can be heard, and that involves a heavy moral responsibility.

Futures critical: a review

John McHale

My central purpose will be to raise questions rather than to provide answers: to be abrasive rather than reassuring: and to try, as far as possible, to push our thinking out of the somewhat well-worn tracks of much current debate.

There seems to be a general agreement that we live in a period of more critical transition, of revolution, and of more abrupt discontinuity in human affairs than ever previously.

Even the phenomena of change, in themselves, differ from the past. We are no longer dealing with isolated sequences of events, separable in time, in the numbers of people affected and the processes perturbed. There is a vastly greater simultaneity of occurrence, swifter interpenetration, and increased feedback of one set of changes upon another.

Though the origins of many of our changes may be located more remotely in past developments, two critical aspects have become dramatically visible within our own lifetimes.

- One is the explosive growth in our actual and potential capacities to intervene in the larger processes which govern our collective survival. Global in scale, capable of affecting the physical balance of life on the planet itself, and reaching into every aspect of individual human life and societal institution, our ongoing change patterns now constitute a social and ecological transformation of unprecedented magnitude. It may indeed be hypothesised that in three to four generations we have begun to approach an *evolutionary* magnitude of change within a *historical* time frame. This is evident in the explosive growth of human numbers; the transformation of the earth to human purpose; the range and amounts of energies and materials exploited; the possible transition from marginal survival to potential abundance — and in the degree to which many of our previously local problems are now writ large on the world scale. In effect, the scale of these changes has already altered many of the ground rules which have hitherto governed and defined the human condition.
- On the other hand, there is a severe lag in the conceptual grasp of this transformation, and in the cognitive and affective understanding of the processes through which we may manage its changes more humanely and more effectively.

Both this conceptual grasp of the rate and magnitude of ongoing changes and the control of their longer range consequences have emerged as a major social need.

We have reached a stage in human global development at which continuous review and assessment of the future implications of our present collective actions become crucially important for the survival of human society.

A central premise within this is that we live in an increasingly less deterministic world than at any other historical period.

Human choice, at both the individual and collective level, may now play a major role in determining events and trends — rather than being determined by them.

We are now, more than ever before, in charge of our own destiny.

The most critical questions

Guernier poses these in terms of "The Great Imbalances", ie, of the imbalance of the forces of destruction, of political fragmentation, and of development inequities.

He identifies the dominant problem as that of population and examines this from both the quantitative and qualitative aspects. First, the *quantitative* — that even with the "braking" of population increase, the effects will not be felt till the year 2000 at which point we will have a world population of over 7000 million.

He notes in *qualitative* terms that this will not only be a doubling of the number of people but — as it is the poorest countries which have the highest rates of population growth, the majority of the increased numbers will be underfed and psychologically deficient, both from the nutritional and cultural imbalances.

Looking at the numbers first, one cannot fault Guernier's figures. Our most recent projections, however, show a decline in overall growth beginning in 1980 and continuing through 2000.[1]

This "plateauing out" of population expansion, if encouraged even in current terms, could push the doubling effect to 7·5 thousand million up to the year 2020 — with a decline in percentage growth rate from our present rate 1·8 per year to 1·2 in 2010 and 1·1 in 2020.

But this still leaves the qualitative aspect. The major increases, and persistently high growth rates, will still be in those regions least able to cope with greater numbers. The age differential in regional populations will be critical — with the poorer countries having "younger" populations with most increase in the working age range: the richer with *relatively* older population profiles. The gap between the affluent and poor regions could be greatly intensified.

Our generalised study picture remains, that even if world population growth is arrested, *growth will remain highest in those countries:*
(i)   Least advanced in material socio-economic terms.
(ii)   Having a high population to useable resource ratio.
(iii) Low nutrition, health, and life expectancy.
(iv)   Low individual social expectations and security.

Growth will be lowest (and most stable) in those countries with a reverse profile.

This indicates, however, that the Malthusian association of diminishing supplies of energy, food and material resources with high population is somewhat untenable. The largest population growth is in those countries using the least resources, having least food etc. The affluent countries with declining populations are those using the major share of the world's energy, food and material resources.

We see the only long-term prospect for reducing population pressures in the poorer countries as dependent upon improving their material standards and generalised expectations.

Guernier poses food supply as one of the limitations to this thesis in that a required four fold increase of food production in thirty years is literally impossible for various reasons: past yield records, available land, lack of trained personnel etc.

Here we would differ from Guernier. In terms of actual food supply produced, at the world level, our recent study[2] suggests that there has been no *gross* shortage for some time. For example, world protein production, from both vegetable and animal sources, in the past decade has been estimated at over twice the world requirement. In animal protein alone (excluding fish), average production per year in the late 1960s was around 22 million metric tons allowing for a possible 18 grams per person — more than twice the recommended daily minimum.

This is not to deny the continuing critical aspects of world food where, for example, at the present time, crop failures due to weather and other factors have caused significant shortages, or diet changes have increased demand.

Arable land is not an immediate limiting factor, except in local terms. In terms of maximum yield efficiency, one quarter of the world's arable land is estimated to supply the needs of over four times the present world population.

The questions lie more clearly with the imbalance of production and distribution. Less than one third of the world's people consumes more than it needs; the rest of the world is close to subsistence and perennial shortage. The average person in the advanced regions consumes approximately four pounds of food per day compared with less than one and a half pounds in the poorer regions. The larger amount is also better balanced in terms of animal and cereal products. The North American diet includes 25% of livestock products, the European 17%, the Asian only 3%.

A singular irony in this imbalance is that much of the animal protein providing the richer diet of the advanced countries is produced via livestock feeds imported from those lesser developed countries most deficient in animal protein.

The answers to the world food problem may not be confined to increased local yields, improved agricultural technologies, new sources of food, etc, but lie ultimately within the larger socio-economic and political context of overall world development.

When tied to the "population problem", a further irony is that we need more food, not because of more people — but to have less people! It has become evident, in the past half century particularly, that the growth rate of population declines with the attainment of higher living standards, including adequate nutrition.

The more critical and less directly manageable aspects of the food/population problem are related to water supply and artificial fertilizer.

Though most water use is of a multipurpose cycling nature, the sheer increase in water demand begins to strain the natural return and replenishment cycles. Though agriculture still accounts for about half the water used, vastly increased industrial use and urban growth concentration contributes to the overall supply problem.

Artificial fertilizer demand, with the spread of mechanised agriculture and the development of higher yield cereal grains and other crops, is also a critical factor which may be limiting on the possibility of increasing food yield through conventional agriculture.

Returning to Guernier, he suggests that the demographic barrier alone renders increased food supply as impossible, noting also ". . . the feeling of alienation which is experienced by the peoples of the Third World faced by the development of our modern (ie, technical) civilisation".

The "alienation from a technical civilisation" certainly begs some questions – *one* about the alienation in itself and *two* as to whether there is such a thing as a non-technical civilisation, but discussion of this theme lies more obviously with Mumford's paper which we deal with later.

Guernier favours the call for the definition of a new ethic as one of the major solutions coupled with the creation of new world regional communities and increased aid allocations.

Leaving the ethics theme aside for the moment, one might certainly agree with the need for a larger regional community strategy. This was particularly focussed upon by Menke-Glückert at the first Oslo conference in 1967:

> Because of the high degree of autonomy, more than 140 governments follow their own economic policy and do their own planning with inadequate means at the level of the nation-state or even some lower level. This leads to unadjusted imbalances which grow faster than the means and capacity to stabilise them.[3]

Guernier's argument about regionalism, however, omits the significant growth of the multinational corporate entities. At present –

> Of the 100 largest economic entities in the world, 58 are nation-states . . . and 42 are corporations . . . Many decisions once considered the province of the nation-state are now being made by externally based multinational corporations . . . These decisions may determine whether a country's resources will be developed or not. They may well affect a country's level of employment, balance of payments and rate of economic growth.[4]

Other analysts have suggested that within thirty years such corporate rather than sovereign entities could own about two thirds of the world's fixed assets.

The question patently is whether these corporate entities are extensions of Mumford's "Megamachine" or do they represent a trend towards a more unified "World Without Borders"? Though they may well spearhead the necessary internationalisation of basic economic activities, their current postures leave them somewhat suspect as a wholly positive force in world development.

The more obvious route for the poorer countries may be to increase their bargaining position as *the* resource-rich blocs – with the increased demand for their fuels and other raw materials, they could then be in a position to assert their economic independence as well as gaining a larger voice in the conduct of world affairs.

Both Guernier (and others) take us further into *the overall question of conservation and growth* as reflected in the familiar exponential patterns. Population, resource use, pollution, urban growth, patterns of social violence and destruction are all tied in as exhibiting the exponential phenomenon of doubling or tripling in short time spans.

But our work in this area suggests that growth, size, and change are also relative measures. What looks like separate increases of abnormally explosive growth in one frame of reference may be a more slowly changing distribution in a larger context.

There is a tendency to identify and relate together only those exponentials which fit the case. The growth rate of the discovery of the physical elements is clearly an exponential curve achieving its steepest slope in the past fifty years. Is this a "good" or a "bad" exponential?

We need to remind ourselves also that adequate statistical compilations are of relatively recent origin and still crudely approximate. Many alarming curves, as in the case of some physical and mental disorders, may only be evidence of improved diagnosis and better book-keeping. Social "problems" and pathologies are as you label and count them.

Similarly our perceptual and semantic bias tends to regard technological and industrial activities as the sole agencies injecting vast quantities of "alien" substances into an otherwise perfect system. Forest fires, volcanic ashes, locust swarms, marsh effluents etc are also environmental pollutants (from another viewpoint) – often at a larger scale, and with more severe effects than those of human origin.

This is *not* a plea for unbridled technological optimism – or opportunism; it seeks, merely, to redress the balance of our thinking.

The severe problems of local and worldwide industrial pollution are due to lack of foresight, institutional mismanagement and inadequate planning. They cannot be banished by equally fuzzy and misleading exhortations to return "organic models", "to get in tune with nature" or by unspecified demands to limit growth.

This applies to many of our environmental, socio-economic, urban and technical crises. They are *institutional crises* deriving from specific modes of economic, social and industrial practice which have been encouraged as convenient and profitable in the short run, and whose consequent long-run effects have now become more visible.

Where this is recognised, the action avoidance mechanism tends to express it as a "crisis of values or ethics". The debate is internalised, and responsibility shifted on to a level where it becomes too personal and at the same time too universalistic to be approached by more practical and rational means.

At one end, it vanishes into the limbo expressed in "cleaning up your personal ecology", and at the other, it is viewed as an inevitable contradiction within "Western Civilisation", or as part of the larger "runaway world syndrome".

In some cases, the rush to protect the future from the consequences of an improvident present looks less like concern with the future than with a nostalgic return to past (and even) pastoral simplicities. Despite the sophistication and rigour of the "growth limits" and "resource exhaustion" schools, their arguments share, in varying degree, an intrinsically conservative set of latent premises.

We have gone too far, too fast, perhaps even too comfortably! The dream of potential affluence and abundance is over, and the "iron law" of the exponentials will punish our hubris by forcing us back upon simpler and more straitened circumstances.

Obviously, there is much that one may agree with in these recent assessments of possible catastrophe, if many of our current malpractices are not corrected. On the other hand, the too ready acceptance by the various "official" establishments of the neo-Malthusian constraints and limitations on further material improvement of the less fortunate is, in itself, somewhat suspicious!

One cannot deal adequately with the "diminishing resources" argument in this brief review. But a few comments from our own work may be in order here.[5] We have examined the two major lines of debate:

- The growing depletion of resource reserves as a result of hyperactive consumption in the past hundred years.

- How this is associated with continued increase of population and living standards to suggest:
    (i)   Curtailment of future growth through material shortages.
    (ii)   Rising costs in extraction and use of lower grade ores.
    (iii) Increased environmental deterioration as contingent upon increased materials' usage.

First the growing depletion argument ignores the fact that materials are not "used up" or depleted at the envisaged rate. They are used for various purposes in various periods and are then "discarded" and/or re-used. Even with existing disincentives we are currently recycling materials at an increased rate.

Material reserves are conceptually and technically dependent — and usually thought of as resources in the ground. The cumulative above ground reserves are not accounted for, eg, we found that calculating 100 years of US production and use (even given wastage) provided a respectable percentage reserve figure for many key metals and minerals in use or dump stockpiles.

Depletion is also offset by the rising rate of extracted performance per resource unit used. We tend to use less material per product or function — even with little or no direct policy attention. This is, of course, more marked in advanced technological production.

Rising material costs were not found to be associated with increased extraction/ processing costs. Labour, capital, energy costs have been decreasing over time in most technically advanced production and service sectors.

Similarly, there is no absolute relation between increased production and consumption, and environmental deterioration. Some of the richer industrial countries have significantly "cleaner" environs than the poorer ones. This is an institutional and regulatory problem rather than an ideological or moral one.

In relation to projected population increase, it is noteworthy that major increase will be in the lesser developed regions whose material resource usage is relatively negligible — whereas most anxiety is felt in the developed regions whose populations are stabilising and whose consumption of certain materials is levelling off!

The problem again is one of imbalance. Approximately 75% of major industrial materials are consumed in the advanced countries; 25% in the poorer two thirds of the world. But this position, though inequable, is far from stable: even in the hyperactive economies there is the likelihood of a slackening demand for raw materials in the longer range as their populations stabilise and as their production sectors become a declining proportion of their overall economies.

The more central cause for alarm lies *not in the shortage* of materials but in their increased availability – as encouraging policies of hyperactive and wasteful consumption.

When we turn to energy the picture is somewhat similar in its maldistribution. The USA with 200 million people uses more energy than the UK, USSR, Japan and Germany combined, for over twice the number of people.

But we may note again that population increase is unrelated to the so-called energy crisis – as the latter is most evident only in those countries with low population growth.

Can we tie increased energy consumption directly to increased GNP? Beyond a certain level of economic and technological development even this is not the case! In the USA, for example, there has been a steady decline in energy input per dollar GNP over the past 30 years – the increased demands have not been in the production sector but in transportation, services and residential uses – characterised by overstimulation of demand and creation of relatively artificial high energy needs.

Actual requirements for sustaining the same or higher living standards could be much less.

In terms of the future, the world energy crisis is one whose outcome is dependent not on the overall availability of energy sources whether fossil fuels, nuclear and other energies, but on conventional institutional, market practices and vested interests and the lack of any coordinate long-range policies for preferred uses of fuels, and for adequate levels of research and development into alternative energy sources and more rational modes of consumption.

Concepts of a return

It seems appropriate, at this point, to turn to Mumford's paper on "technics and culture". I must confess here that critical comment is tempered by one's debt to his pioneering scholarship in this area and by the extent to which he has illumined all of our views.

His present paper is particularly cogent in reminding us that:

> . . . much of our thinking today in technocratic circles is being done by one-generation minds . . . that many of our current technological and other developments had significant origins in periods prior to the Industrial Revolution . . . (eg) . . . the invention of the mechanical clock in the 14th century did more to advance modern technics than the steam engine or the automatic loom.

One might also underline that the "one-generation" mind is certainly prevalent in much futures thinking where it specifically neglects the consideration that many of our present discontinuities are the "cresting" of longer wages of developmental change rather than uniquely present in our period.

This is not to suggest the inevitability of historicism. It recognises rather that a prime motto for futures research might be Toynbee's assertion that, in terms of overall human development, "all recorded history is contemporary history".

The core of Mumford's argument is that whilst asserting the primacy of various technological elements in human development, he views with alarm the recent ascendancy of mechanisation over humanisation via the megamachine – that the meta-technology of power concentrated in institutional structures forces man to conform to the systems' need rather than his own.

Whilst one would certainly agree with the location of major crisis and dys-function in the overweening autonomy of institutional power, there are several points in his thesis which beg certain questions.

Firstly, as Mumford himself points out, the meta-technology of over-constraining organisational structure is not unique to our period. He locates its origins in the urban revolution of the earliest civilisations . . . "When large populations were first organised and put to work on a scale never before conceivable . . . and with a machine-like precision and perfection never possible before".

Secondly, whilst acknowledging that "Man is his own supreme artifact . . . and that every form of technics has its seat in the human organism", Mumford then falls back on the somewhat artificial dichotomy of an "inorganic" and alien technology in conflict with the "organic . . . polytechnics of the handicrafts – pottery making, spinning, weaving, stone carving, gardening, farming, animal breeding . . . the rich repository of well-tested knowledge and practical experience". This he typically enshrines in the small community as "transmitting the essential traditions of work, aesthetic mastery and moral responsibility". This organic model is associated with what he calls "the rural factor of safety".

The recurrence of this *gemeinschaft* model of the organic solidarity of the craft community echoes Weber's phrase, "that the fate of our times is characterised by rationalisation, intellectualisation and above all by the disenchantment of the world".[6] This estrangement from modern society has been a major preoccupation in the social sciences since the outset of the Industrial Revolution. It is still an enduring element in much futures thinking, both in the conservative and counter culture wings.

One might locate some of its origins in the rise of "indeterminacy and uncer-tainty" in the 19th and early 20th century. From being a relatively contained, fixed, and "rationally" apprehendable Newtonian world the whole order of reality began to shift its outlines, become ambiguous and infused with relationships which were neither visibly nor logically apparent before. Alice not only voyaged into Wonderland, but went through the Looking Glass.

It is here, I think, that the author of "The Myth of the Machine" falls victim to the machinery of the myth!

There is indeed little to suggest that a return to the homogeneity, uniformity and social constraints of the pre-industrial peasant enclave would provide more freedom, more humane life-enhancing opportunities etc, than modern urban living – even, perhaps, at its worst.

If anything, despite the supposedly widespread alienation and anomie, the more fortunate individual today is less constrained, more self-consciously aware and more concerned with the quality of life than at any time in history.

Where much romanticised by those who did not have to endure it, the golden age of the peasant was hardly even gilt. Before the factory, women and children worked in the fields from dawn till dusk. The tenant craft-farmer is no less bound and disciplined by relation to the land than by machine tending.

Though supportive of humanitarian concern, the implied return to some pre-technological, tension-less and stress-less Eden, to the earlier simplicities and securities of a more "natural" order, is merely a recurrent element in our mythology.

The image of the more pastoral society as conserver and guardian of nature is also historically false. Both the successive herding and agricultural phases of our development had, in their turn, relatively enormous impacts on the natural environment. The effects of the re-ordering of the balance of animal species, the overgrazing of land, the substitution of monocultures for diverse wild strains, the development of elaborate irrigation systems etc are still visible.

By sustaining these uneasy dichotomies between "the natural" and "the artificial", we block the way towards a more rigorous reconceptualisation of the need to restructure human society. One may re-emphasise the *integral* quality of human activities by capping Mumford's assertion that "Man is his own supreme artifact", with the axiom that human artifice is the natural order for human beings!

Whilst we do need to control that "artifice" in its more negative predilections, one cannot agree with Mumford's statement, "If our main problem today turns out to be that of controlling technological irrationalism, it should be obvious that no answer can come from technology".

Surely it is the evidence produced by greater scientific and technological understanding which has led to the identification of many of the side effects of technology, and to our wider possibilities for controlling the negative aspects of a technology before its use becomes widespread.

This type of anticipatory technological assessment is perhaps best exemplified in the directions of large-scale modelling which are reviewed by Cole and Sinclair.

Simulation of the future

They underline that "At the lowest level, all formal models help a researcher to clarify ideas through forcing a resolution of inconsistencies . . . in theories evolved often for different purposes and based upon data gathered independently".

At a higher level, one may suggest, of course, that the central utility of computer modelling is that very large-scale interactions can be simulated in small-scale analogues. Processes and events which might take weeks, months or years to occur in real time may be run through in a few hours or days — and, importantly, various effects and hazards can be determined with no "real life" costs.

Cole and Sinclair reiterate, however, "With regard to the numerical aspects of computer calculation (that) there is a danger in treating these complex tools as 'black boxes' robust to all situations and utterly dependent and accurate. They simply are not".

The use of such predictive simulations to direct policy, in the larger sense, is another matter. As Cole and Sinclair suggest, ". . . their role is not to supplant all other inputs into the decision process but rather to supplement the various perspectives brought to bear on practical policy issues".

There is another way, of course, in which the results of such models can, and do, influence policy and public opinion; that is, *the unintended consequences of pro-*

viding new sets of social metaphors. This has been particularly noticeable in "The Limits to Growth" which had a sharp impact on all sectors of the society, from the policy makers to the man in the street – much less through the rigour of its modelling than by its compact visual and poetic metaphors on the phenomenon of growth and decay.

Some of the dangers inherent in large-scale world modelling of this kind are rather obvious:

1. The selective personal screens and latent assumptions which may not only bias the theoretical structure of the models but the kinds of information which are admitted for consideration. It is easiest to use both models and data which are 'administratively convenient' eg, those most amenable to quantitative measure.

2. Again, as Cole and Sinclair note, much statistical data at the international level, ". . . reflects a mosaic pattern where each piece has been gathered for a different purpose and in a form which reflects a need peculiar to the theoretical underlay of a national culture and the statistical purposes of the data gathering organisation".

3. One would advance also the tendency to reify such models, that is to assume that the model *is* the world – rather like mistaking the highly abstracted map for the actual physical terrain.

4. Much has also been made about the *counter-intuitive* power of such models, eg, Jay Forrester's assertion that:

> Evolutionary processes have not given us the mental skill needed to properly interpret the dynamic behaviour of the systems of which we have now become a part . . . (computer) models can now be constructed that are far superior to the intuitive models on which we are currently basing our future survival.

However, intuition may be described as the integration of *weak* signals, a function which may be more crucially important to the understanding of complex systems than Forrester would allow. The counter-intuitive superiority of computer models might also be faulted by Ashby's "Law of Requisite Variety" which notes that variety and complexity in a system may only be controlled (or understood) by a system of the same, or greater, level of variety.

In considering the place of world modelling in policy making, the authors are somewhat less sanguine in their assessment. They view the policy process as necessarily involved at each stage with political and administration questions regarding the costs and benefits to varied constituencies. Though susceptible to methodical analysis it is, in practice, compounded of a delicate balance of goals and preferences, estimations of possibility and the negotiated outcomes of interest bargaining. They note particularly that "world models should, but do not, take into account the time span and behavioural uncertainties of this process".

The key point in their critique is the weakness in accounting for the *social context* of decision making:

> World modellers appear to have no more success in finding the lasting and important elements of social structure, and hence social change, than have other less ambitious analysts. Indeed current attempts at world modelling have been almost startlingly bare of any attempt to incorporate enough the modicum of material that is now known about human behaviour in particular situations of interaction.

They follow this point up in discussing the need for a historical perspective in such modelling but, whilst noting that such historicism may provide a comforting "feeling" of continuity,

> The constraints of present action arise not from the inevitable flow of historical trends but from the views that men hold in their heads of what history implies and neither can these be described in material categories.

Again, if a socio-historical stance is taken,

> . . . then the predictions made must be viewed by decision makers in the light of the nature of the stance. A Malthusian base will give Malthusian answers, and only the timescale can be claimed to be particularly pertinent.

Their central concern is about the dubious authority of such models despite their apparent precision and numerical sophistication.

> In the current dearth of scientific knowledge of the future there lies a danger which must be avoided at all costs, that is the attachment of spurious certainty to predictions of tenuous truthfulness.

The authors offer a much more detailed assessment of the Forrester type model than we can review here, and conclude that, "whether or not confidence can be placed in (its) actual results, it provided a stimulus to a large amount of work on large-scale modelling which aimed to examine a variety of world problems".

In surveying a number of other such models in their paper, Cole and Sinclair remain ". . . rather sceptical of the immediate usefulness of large-scale models for greatly assisting the formation of public policies . . . (and are) not convinced that present models, despite their formality and precision, approach the standards required".

They note ". . . two paramount issues to be tackled before these models will realise their full potential. The first is the establishing of a satisfactory comprehensive and consistent knowledge base. The second is the satisfactory integration of models (both their development and results) into policy-making decisions".

In looking at the future of such world models, Cole and Sinclair are a little more sanguine about their long-range possibilities, ie, "even if not for their usable results but for the ways in which we should use them to seek out worthwhile goals".

This note regarding "goals" brings up the point that the development and integration of such models into systems for more efficient planning and regulatory management may tend to reinforce institutional determinism as against the individual balance of influence in society. Vickers has addressed this point in noting the lack of comparable *appreciative systems* which can generate ongoing review and criticism regarding changing goals and values.

> Men, institutions and societies learn to want as well as how to get, what to be as well as what to do; and the two forms of adaptation are closely connected. Since our ideas of regulation were formed in relation to norms which are deemed to be given, they need to be reconsidered in relation to norms which change with the effort made to pursue them.[7]

Goals and value preference: the theme of development and new societies

As may be noted in passing, Mumford tends to oppose two developmental models; the inorganic, or high technology based path – to the organic – functioning at a lower level of technological constraint and more directly related to craft-based "close to the land" communities. Guernier inverts this in stressing that the imbalance in world development lies in the fact that too many people are still bound to the organic rural/craft model. He suggests in effect that it is only the richer nations which are developing; the rest are stagnating.

One might agree more, in part, with the latter view. Even the so-called advanced nations are still undergoing a painful "three generation" transition into modern societies, and are marked by severe dislocation, deterioration and obsolescence in critical areas of their socio-economic and political structures. Many of their internal institutions are archaic, strained towards breakdown and confined by obsolete concepts and practices. Their physical environs are still suffering from the backlash of the initial developmental phases of unrestrained industrial exploitation. Though we refer glibly to Western scientific and technological societies, none of them have yet approached what might be termed a scientific society, ie, one whose motivations, goals, and orientations are congruent with, and permeated by, the scientific outlook in the larger sense.

Dator in his paper avoids both of the development viewpoints — by rephrasing the given question,

Is the social model of development in East and West necessary and still valid within certain limits for the countries of the South?

He asks rather,

Is there a crisis in the image of the future which is signified by the term 'development', such that the image has lost much of its appeal to persons in developed and developing countries alike. If so, why? And if so, what might replace it?

Then in making the case for his model of "The Transformational Society", ie, beyond development, he indicates that it need be "neither utopian nor dystopian, but eutopian — (neither ideal dream world nor nightmare) — but the image of a feasible desirable place (and time) which is significantly different from, and better than, the present or the past".

Though Dator in his approach to this "transformational model" specifically disavows "analysis and dissective criticism, in favour of synthesis or creative imaging" — a goodly section of his paper is devoted to a closely analytical assessment of both the "Rostow-type" models of the necessary stages and of the opposite camp of "ecology-counter-culture-world-order" and ethical critiques.

Within Dator's analysis there are several questions to which I was particularly drawn, eg, are certain ethical attitudes more conducive to development or not? We have heard much about the Protestant ethic in Western development and its inapplicability to Eastern and Southern cultures as an alien and disruptive conponent. Dator rather neatly torpedoes this point in underlining the functional equivalence of both the Marxist ethic and the Japanese Bushido code in replicating the attitudinal conditions of the Protestant ethic for a number of non-Western situations.

As an aside here, one may note that the Japanese 17th century Tokugawa "Edict of the Hundred Articles" could have been co-authored by Calvin or Ben Franklin from its Puritanical cast, eg, "Avoid things that you like and turn your attention to unpleasant duties" or "When desires arise in your heart, think back to the times when you suffered distress".[8]

To those who might still blame the Judeo-Christian value clusters surrounding development as inimical to those Eastern philosophies conventionally revered for their organic harmony with nature, Dator indicates that the latter encouraged as much environmental modification.

He quotes, to effect, studies by Yi-Fu Tuan and others detailing some of the ecological disharmony occasioned by non-Western religious and cultural values, eg,

- Buddhist cremation practices as sufficiently widespread during the 10th to 14th centuries to create a timber shortage in SW China;
- in 17th century Japan, temple building was responsible for seven tenths of the marked deforestation;
- in the most civilised of the arts — writing; soot was needed for black ink, and soot came from burnt pine. So even before T'ang times, the ancient pines of Shan-tung had been reduced to carbon, and now the brushes of the T'ang dynasty were rapidly bringing baldness to the mountains . . . In modern Shantung, one conspicuous strip of remaining forest owes its existence not to native piety but to some conservation-minded Germans;
- on recent reports of new growths on the once bare hills of South China, it must again seem ironic that this "mist of new green" is no reflection of the traditional virtues of Taoism and Buddhism; on the contrary it rests on their explicit denial.[9]

On the human side, one might add the footnote: the exposure or sale of female children, widow "suttee" or the assassin cults as not atypical features of non-Western social harmony.

I have dwelt overlong on these points because of the current fashion of the Western disruption theme, of the traumatic effect of progress and development on har-monious traditional cultures. Again, this is not to say that all things are holy in the name of progress, nor to suggest that we should wantonly destroy traditional ways of life by unthinking technocratic means. It is noteworthy, however, that the notion of cultural barriers to the poorer nations' "readiness" to use Western tech-nology usually applies more to the raising of their living standards and expectations. It is rarely invoked in the case of military development. The individual may go from a spear to an automatic rifle in a few days without any appreciable cultural trauma!

We should also recall that much of what we describe as Western scientific and technological development had its substantive origins in the East. Can we reason-ably talk about a specifically African physics with a different set of natural principles, or a European biology, or an Asian chemistry?

Returning to Dator's thesis, whether or not one agrees with his concept of "The Transformational Society", his approach to this theme is very important for this discussion.

He does not start with the description of the Society but with "its values, goals and human rights". His basic considerations of these then flow from "the possibility or desirability of making human freedom, within a society of abundance, the prime goal of a future society".

On the way to such a society he believes "that we can handle much more change than we are currently experiencing – biologically, psychologically, culturally and environmentally – and that we can and should work for the creation of a world which maximises our ability to live even more humanely in a situation of rapid change".

Whilst reiterating that his design is "beyond development", Dator underlines agreement with Goulet's definition in "The Cruel Choice", quoted as follows:

> Development is a particular constellation of means for obtaining a better life . . . (with) . . . the following objectives:
>
> - to provide more and better life sustaining goods to members of societies;
> - to create or improve material conditions of life; related in some way to a perceived need for esteem; and,
> - to free men from servitudes (to nature, to ignorance, to other men, to institutions, to beliefs) considered oppressive. The aim here is to release men from the bondage of these servitudes and/or to heighten their opportunities for self actualisation, however conceived".[10]

Dator inverts these in his scheme of priorities. Whilst agreeing on the basic material needs, he places greatest emphasis on individual freedom, ie, "a society in which every human is totally free in every way from unwanted controls by society as a whole or in part". The dominant unit is the individual – within a minimal set of social and environmental responsibilities.

Though carefully eschewing any ideological label for his design, I think that much of what Dator is saying here belongs squarely within the anarcho-syndicalist tradition. This is not a criticism! The latter is a respectable if somewhat unfashionable body of theories – with many neglected but successful applications in practice.

Besides outlining the individual and social freedoms of his society, Dator also counters in some detail the objections to its viability, eg,

- It is contrary to history.
- It is contrary to human nature.
- It is too atomistic and ethocentric.
- It is too violent and disorderly at best.
- We do not have enough abundance to achieve a transformational society.

The weakest part of his argument, and I suspect that which may occupy the attention most, is the "lack of abundance" objection. Though Dator refers to new types of post-industrial socio-economic systems and appropriate new technologies through which nations may bypass the traditional stages of technological growth, he does not detail these out.

Alternative technological directions

It might be useful, therefore, at this point to present our own summary notion of how the changing profile of technologies might sustain this argument more.

In most discussions of this nature, many of the features considered as enduring and limiting characteristics of industrial technology are actually end phases and adaptations of the older craft production tradition – the factory system, the routinisation of work, the concentration of populations and production centres in close association with raw materials and energy sources. The later evolution of the industrial revolution towards the end of the 19th century furnished the first generation heavy industry, based on the fossil fuels. These still constitute most of our industrial network today – steel, railroads, automobiles, electrical generation from steam etc.

The second generation of industrial technologies emerges most clearly after World War 2 as closely interwoven with nuclear energy and electronic technologies based on new uses of larger areas of the electromagnetic spectrum. Their characteristic profile includes light metals, plastics, computers, aerospace development etc. One significant shift occurs in this generation, towards the more central position of information (or organised knowledge) as the unique resource. Information as basic resource has certain special properties – all other resources are ultimately dependent upon information for their recognition, evaluation and utilisation. It is not reduced or lessened by wider use, or sharing, like other material resources, rather it gains in the process.

> The truth, the central stupendous truth, about developed economies today is that they can have – in anything but the shortest run – the kind and scale of resources they decide to have . . . It is no longer resources which limit decisions. It is the decisions that make the resources. This is the fundamental revolutionary change – perhaps the most revolutionary mankind has ever known.[11]

In this specific sense, the emergence of information and information technologies as the key resource changes the old pre-industrial and industrial zero sum game condition of society into what has been termed the post-industrial phase – into a non-zero sum situation, in which all may potentially win.

We are now entering the third generation phase of industrial technologies whose beginnings bear little resemblance to the first phase. For example, the field of industrial microbiology – the wider and more sophisticated use of microbial populations to produce energy, food, process materials etc; the extension of mariculture; bionic engineering with its direct application of biological principles into new technological forms; the renewed interest in the use of insect and other organic populations for productive work.[12] These alternative "soft technology" tracks are accorded little attention due to emphasis on heavy machine industry as the conventional image.

We may note one signal distinction between these old and new forms. The later patterns of industrial development tend to be relatively non-resource depletive, with lower environmental impact than the older forms. Rather than technological growth necessarily becoming more material it may take place in more immaterial forms – requiring less energy, less physical resource input with correspondingly less residual wastes and environmental impacts.

One could certainly envisage the newer alternative technological directions as providing a basis for abundance that would make the transformational society model more viable.

Search for ethics

Dator poses two basic ethical questions drawn from Goulet's work, "What *kind* of development is human: and *how* must such development be obtained?"

One may recall with profit here the statement of Raymond Williams in a recent BBC article:

> Man cannot derive lessons and laws from the processes of what he sees as a separated nature, lessons and laws supposed to be conditions of himself, conditions to which he must conform. The whole perspective of a man learning from a separately observed nature is deeply false. The correlative is that in the end it is best if we discuss the problems of social and human relationships in directly social and personal terms.[13]

This discussion is the focal point of Simon's paper on the relationship between man and society. His basic premise complements Williams' above, ie, that social conflicts, human needs etc ". . . can only be understood in terms of concrete social contexts and specific human experiences and not in terms of abstract conceptions of Man and Society".

Simon begins, therefore, by posing several questions which probe deeply into a whole range of unexamined assumptions in futures thinking.

His initial rhetorical question is, "Why must humankind survive?"

This wedge is then used to break open two substantive questions which go roughly as follows:

One. What is it that commits us to the future? He notes that ". . . for most of human history and possibly for most of humanity alive at this moment the question in this absolute form was never asked; life was too much tied to the implicit order and periodic disorder of nature."

Two. "Why must humanity survive in ways that we would recognise as being close to our sense of the human?"

We might cavil at the way in which the first question is stated. Some sense of, and commitment to "the future" seems to be relatively constant in all human cultures — even though it might take many forms, eg, of cyclical return to origins, of predestination, reincarnation etc. Certainly, however, the sense of the future changes with the variable nature of human consciousness and is not an absolute form.

For most people, most of the time, survival in the present over-rode preoccupation with future states. One might still suggest that our contemporary sense of the future has emerged most strongly in those fortunate to live in conditions beyond such marginal survival constraints.

Much of that "sense of the future", however, is still bound up in a "problem-oriented" view of modern society which reflects an underlying pessimism and troubled uncertainty. It is still overhung with early social science reactions to the disruption of traditional forms and the social disorder attendant upon rapid change. Thus, the inbuilt limitations on human experience, the inevitable constraints of society upon the individual, the need for collective security and the cohesion of traditional bonds to escape alienation and anomie, the routinisation, and psychic decay of mass society etc, are still the major stated and unstated premises of much futures thought.

This *value indictment* of contemporary society, though supportive of important humanitarian concern and ameliorative social action contains many mythical images of 'man and society' which may well hinder more radical exploration of social change and limit our expectation of human options.

Simon's second question on the definition "of the human" deals with many aspects of this and strikes directly to the heart of our discussion theme. He notes that "we must begin with a sense of change potential in the most critical of all variables — man himself".

As we go forward in our discussions we will hear repeatedly the terms — alienation, anomie, optimum development, human actualisation and potential etc.

Lurking behind the labels are various models of man which lay claim to all sorts of conditions of universality and historical continuity — linked to implicit assumptions regarding man and nature, human needs, social health and pathology, and so on.

Simon dismisses many of these cherished notions in suggesting that:

(1) The various universal models that we are offered . . . tend to be celebrations of an image of the human with which we feel most comfortable.

(2) Conceptions of man's universal nature undoubtedly give rise to attractive politics, eg, the brotherhood of man, but also tend to give rise to a disguised ethnocentrism and consequent conservatism. We learn to live with a protean conception of social life by retreating to a seemingly fixed conception of the human.

(3) Implicit agendas for 'optimum development' rarely pause to ask: Optimum for who? or, for what?

(4) We may have to begin to face the fact that as the social order changes so do people and that these changes occur on the most fundamental levels.

(5) Etc.

In his central discussion of the "fallacy of sociomorphism", Simon underlines that many of our implicit models:

. . . impute to the individual organism that which is properly a function of social life, (and) which allows us to accept uncritically as constants, as sociocultural universals, or 'natural laws', for social life that which may, in fact, represent little more than the outcomes of the sociohistorical moment.

He notes the dangers of this sociomorphic attitude in the persistence of concepts like alienation with their underlying sense of some ultimate fixed human nature that various social contexts repress, or distort. And in the ways in which this attitude overconstrains and rigidifies the sense of being human — not as being in continuing change but as a set of givens which restrict change.

To really accept the idea of a protean (diversely plural) society is also to accept the idea of protean people. Both are inseparable aspects of the same process . . . Our models or scenarios of the future must incorporate the human as a variable in the multiple sense in which that term may be used.

This suggests to me that the ideal of personality development towards a more fixed stable identity (as "normal" and "healthy") is open to question — that indeed identity may be recognised as a profile of attributes which change over the life cycle, as an essentially unfinished nature.

It moves discussion of individual needs and natures away from the images of Marx/Freud as overconstrained by the past; and those of the Skinnerian variety as over-conditioned by the present to one in which identity is a process compounded of past references, present demands and future preferences.

In line with many other commentators, Simon locates the future as post industrial and explicitly links this with a shift from material deprivation and marginal survival to potential affluence. He differs from most, however, in suggesting that what may be strongest in this post-industrial shift is not a looming sense of the future but a developed capacity to enjoy the present!

In many ways, he notes that some of our preoccupations with the future may be linked to a denial of, or discomfort with, the present, and to a postponement of gratification and achievement which is rather like the so-called Protestant ethic — in the sense that present decisions and actions are justified in terms of some future reward.

> The children of the affluent being, as it were, born on the far side of achievement and success, tend increasingly to develop an immunity to the oppressive sense of the future that tyrannised so many of their parents. (Even their parents increasingly require the invention . . . of alternative therapeutic modes to be temporarily reassured of their own reality and meaningfulness. Indeed, futurism, which rapidly takes on many of the forms of a social movement, may function for many as just that kind of therapeutic mode).

In delineating, more specifically, some of the ongoing social shifts towards an emerging post-industrial society of affluence, Simon comments on:

1. *The decline in role stereotyping* — as people are increasingly freed from narrow sex role alternatives and less limited by work and non-work related role fixation, there will be more flexible (and ephemeral) sets of life style and role commitments.

2. *Lessening importance of the family* as central economic and socialisation unit — and *its growing importance* as a context within which personal needs can be developed and expressed. (NB: "Industrial society . . . accused of taking people away from the family, actually increased the amount of time people could spend within the family . . . particularly increased the amount of time not bound up with routine activities but time free to attend to one another's needs".)

3. *Changing conceptions of work and non-work* — the potential diminishing of work per se — "through being less chained to any one job and the weakening of the job relationship to other aspects of life . . . will also pose in equally urgent ways the problem (not only humanising work but) of humanising the non-work aspects of life . . . We have been 'trained' for a long time to work (but) rarely, if ever, been trained to fully live in other ways".

4. *A decline in materialism.* "From a society in which material possessions have often become the measure and justification of an entire lifetime's labour — the very technological capacities of a post-industrial society, producing more and better at lower costs reduces the significance, particularly the moral significance of material things . . . (material possessions) are less likely to serve as powerful motivating factors in their own right".

5. *As the scope of social life expands,* the need and ability to organise life in more personal terms increases. (There is) the potential release of people "from the ultimate totalitarianism that has characterised most of human history . . . to live at ease in a world where there may not be any purpose larger or more compelling than their own.

6. *Simon concludes* on the notion that "Man's capacity to be self-regulating without being socially dangerous (is) not only possible but possibly necessary in order to manage a post-industrial global society".

The central challenge

We have gone from discussion of global imbalance and world models, through resources, technology and culture, to the changing nature of human needs.

Much of our review has underscored the critical dimensions of the problems which face our world — but it also suggests that, to a large extent, we already possess the physical means to surmount the material aspects of most problems.

What we lack, specifically, are the requisite sets of socio-economic and institutional arrangements through which these means may be used to more human, and more humane, advantage. This is the central challenge!

If we are able to meet it, we could not only pursue many alternative visions of new societies — but turn such pursuit into a major social purpose in itself. We could proceed, on a larger and wider scale, to engage with the process of learning to be human, of exploring in myriad ways the diverse modalities and potentialities of the human condition.

References

1. John McHale and Magda Cordell McHale, *The Timetable Project Stage I: An Assessment of Projected Relationships between Population and Resources*, 1972—73. Report to National Science Foundation
2. *Ibid.*
3. P. Menke-Glückert, "Proposals for an International Program for Joint Technological Endeavors for Peaceful Purposes", in *Mankind 2000* ed R. Jungk and J. Galtung, Oslo University Press, 1969
4. Lester R. Brown, *World Without Borders*, New York, Random House, 1972
5. John McHale and Magda Cordell McHale, *The Timetable Project Stage I: An Assessment of Projected Relationships between Population and Resources*, 1972—73
6. *Also*, "Man is freed etc . . . but also from security and a sense of belongingness; he is disenchanted, disillusioned and estranged".
7. Sir Geoffrey Vickers, "The Sociology of Management", 1964
8. G. B. Sansom, *The Western World and Japan* (1949), New York, Vintage Books, Random House, 1973, page 181
9. See references to Yi-Fu Tuan and Schafer in Jim Dator, "Neither there nor then . . .", elsewhere in this book
10. Denis Goulet, *The Cruel Choice: A New Concept in the Theory of Development*
11. U Thant, Address to Students' Association. The Decade of Development, Denmark, May 8, 1962 (UN Press release SG/1194)
12. John McHale and Magda Cordell McHale, *World Trends and Alternative Futures,* East—West Center, University of Hawaii, Honolulu, Hawaii, March 1973
13. Raymond Williams, "Social Darwinism", The Listener. November 23, 1972 (from one of a series of lectures delivered at the Institute of Contemporary Arts, London, entitled "The Limits of Human Nature")

The great imbalance

Maurice Guernier

Today's generation of leaders finds it difficult to grasp the predicament of the modern world and its evolution, because they have experienced both the end of a static and partitioned world and the beginning of a dynamic and planetary world.

During the last century, of course, some so-called industrial countries were already undergoing a development process, which was altogether unintentional; it was said that "the wealth of those nations was developing". But this evolution was slow, and the different social classes were hardly aware that any progress was being made at all.

Meanwhile, all the other regions of the globe were stagnating and, because of the colonial system, making only the smallest advances toward establishing commercial activities. The inequalities in the various countries' standards of living were no less flagrant than they are today, but no one thought of them as being unjust. This was clearly expressed by Mrs Indira Gandhi when she said, "For millennia India has suffered from malnutrition; but what is serious today is that she knows it".

Explosive change

Beginning at a moment that cannot be fixed exactly but which occurred some time after the second World War, the world has been shaken by a series of explosions, which will certainly mark a turning point in the history of humanity. The atomic explosion was the first to herald the new era. Men immediately became conscious of the fragility of the planet they inhabit as it spins in the cosmos. Henceforth, man can destroy humanity. Thus the relatively small frontier problems among nations, the problems of wealth distribution among the social classes, of religion among groups and of races among continents will be more and more dominated by the overwhelming fear of the atom. From that moment henceforth, the equilibrium of terror was to be the crucial element of nations' politics and the super-powers' rivalries.

The second explosion involves the political sphere. A great many nations sprang into existence with the collapse of the colonial empires. When the United Nations was created in 1945, its membership comprised 51 nations; 20 years later this number had risen to 114. In 1960 alone, 20 new nations were born, some of which had no more than 400 000 inhabitants, while China's population numbered over 650 000 000.

The third explosion was that of production. Following American initiative, the capitalist industrial nations — and 20 years later the socialist nations as well — have devised new methods, labelled "productivity", which have increased the process of development at a pace hitherto unknown.

The case of France is particularly significant. In 1950, a long period of stagnation and decline came to an end, immediately followed by a steep rise, achieving levels of performance which never before could have been imagined. Each nation, now rated according to its economic growth, finds itself virtually engaged in a development race.

Finally, the fourth explosion, the population boom, which has occurred simultaneously with the other three and follows the same pattern. It is the most extraordinary fact in humanity's long history: from an estimated population of 250 million at the time of Christ to approximately 500 million in 1650, it had taken about a millennium and a half to double the species. It will take only 35 years to double the 3706 million population in 1971 to reach, according to UN computations, 7400 million in the year 2000.

Such are the great explosions that have struck the world we live in. We shall show that they have been responsible for serious, widespread imbalances, and that each of them constitutes a permanent threat of conflict.

Inherent instabilities

The threat of nuclear war is now generally held to be stabilised, since the power of dissuasion is such that the super-powers, which control thermonuclear weaponry, are somewhat neutralised. By virtue of restraint, they are obliged to maintain peace. This is demonstrated every day throughout the world. But if thermonuclear balance exists among the great powers, there is still a fundamental imbalance between the great atomic powers and all the other nations which do not possess the atomic bomb. They must bow to the super-powers.

With the expansion of technology, however, what will happen if atomic weaponry falls into the hands of the small powers? One hardly dares imagine the effect on the thermonuclear balance of the big powers if some small, beleaguered nation out of desperation should choose to unleash the atom — and if each nation engaged in a conflict should consider itself in a desperate situation. Never before have the forces of destruction been so sharply unbalanced.

The political explosion has also intensified world imbalance. Never before has it been possible for delegates representing a few hundred thousand people to sit at the UN Assembly with the same voting capacity as those who represent a thousandfold more. This dislocation of humanity into a great number of small nations and large blocs is all the more serious in that a revolt by the smallest can embroil the largest, and thus constantly threaten world peace. The Soviet and American giants must confront the dangers inherent in the problems of a million Palestinians, of the Biafrans and the South Vietnamese.

With the development explosion, the world imbalance among countries is still sharper. The figures are well known; we need only mention a few:

An American farmer feeds . . . . 45 persons
A French farmer feeds . . . . 12 persons
A Russian farmer feeds . . . . 5 persons

```
A Senegalese farmer feeds . . . . . . . .    2 persons

In industry, where a Third World worker produces    1 unit
                   the European produces .  .  .   20 units
             and the American produces .  .  .   40 units
```

The countries which are actually in the process of development are the industrial nations of the capitalist and socialist blocs. The Third World countries must be qualified – upon the evidence – as stagnating; the gap between these two blocs widens unremittingly from year to year.

Population

The demographic explosion, by its consequences, will determine a series of fundamental imbalances over the next decades. The global population expansion is, and will remain, the world's dominant problem, at least for a century.

We must examine it from two different angles: quantitative and qualitative.

From the *quantitative angle*, one certainty can be assumed: that the present population will approximately double by the year 2000. Even if brakes are applied, the effects will begin to be felt no sooner than the year 2000. We are certain therefore that the world's population will exceed the 7 thousand million mark and, as we shall see, will inevitably reach 10 to 12 thousand million during the 21st century. Are there any limits to this growth?

Obviously yes. We know that, at the present rate, by the year 3000, only about four square inches of living space would be available for each man. Still at the current rate, it has been calculated that, by the year 2070, ie in a hundred years:

- France would have 134 million inhabitants (245 inhabitants per square km, as in Germany today)
- The USA would have 420 million
- Russia would have 454 million
- Morocco would have 1505 million (16 million today)
- Bangladesh would have 3869 million (74 million today)
- China would have 20 850 million (750 million today)
- India would have 25 418 million (576 million today)

The soaring figures applied to the countries of the Third World clearly show that exponential growth in biological and limited spheres demand that curbs must be imposed.

What will these brakes on demographic expansion be, and will they be natural? Professor Jonas Salk, of the Institute of Biological Studies, San Diego, California, has studied the natural limitation of growth on the fruit fly *drosophila* in a limited environment. His deduction is an S-shaped curve, ie a growing line such as that of human population which, however, flattens out to become horizontal.

Professor Salk, like the French demographer Alfred Sauvy, estimates that "the fecundity drop has already started in most countries of the world and tends to accentuate henceforth"[1], and that "from all parts of the world is forthcoming manifold evidence which suggests that Epoch B (progressive decrease) has already definitely started".

These two specialists may well be optimistic. Men, in any case, are motivated by factors — religious, sociological, political — which differ from those of the fruit fly!

Many authors, such as Professor Josué de Castro, a Brazilian, and the Greek economist A. Angelopoulos, believe that growth in the GNP will be mainly responsible for birth limitation; indeed a rise in the standard of living has always been followed by a slackening of demographic growth. But we strongly doubt the validity of this argument, mainly because no rise in the standard of living is possible when demographic growth rates range from 2·5 to 3·5%.

One of the predominant causes of world imbalance is the fact that the poor countries claim the highest rate of demographic growth; therefore they are the countries which confront the greatest difficulties in advancing their development and consequently in curtailing the birth rate.

The Chinese were the first to understand that their nation would not develop with a demographic growth rate exceeding 2%. The Indians have also understood this. No other country of the Third World has even begun to tackle this critical problem.

Professor Paul Bairoch[2] has clearly shown that "the demographic inflation represents a most important barrier to economic take-off". He writes, quite correctly, that "In the case of a relatively modest objective of a 2% growth per capita in the GNP, the necessary capital investment (ie the amount to be invested each year in productive equipment) must reach 21% of the total gross product".

This means that the unfortunate Indian who earns $100 to live in the most execrable conditions, must give up $21 so that ten million children born every year (who, before they can work, will join with those too old to work, to constitute an immense inactive and unproductive population of 300 million) may be fed, clothed, housed, educated, equipped, etc. All the Third World countries must absolutely understand that they cannot develop, that they cannot even dream of developing, if the population growth rate surpasses 1 to 1·2% yearly[3]. This is a fundamental choice.

I strongly doubt the value of an authoritarian policy to reduce demographic growth; simple, natural morality forbids it. Rather, I consider it essential that the leaders and élites approach the problem as the Chinese have done, and that each one must refer his responsibilities to the basic factor of all society: the woman. It is she and only she who must decide her family responsibilities with her husband. Finally, only cultural development will help enlighten every individual on his human responsibilities in the matter of procreation.

Quantitatively, the demographic problem is and will remain for a long time a major problem of the planet. Whatever curbs are adopted, the effects will be felt only after the year 2000 and, in any case, the world population could only be stabilised at the approximate level of 10 to 12 thousand million human beings in the course of the 21st century. These figures present gigantic problems on the planetary scale. But the demographic problem is not only quantitative; it is also qualitative and on this level, very serious world imbalances are to be feared.

Nutrition

We know today that a child's nature, his nerve and brain structures, and therefore his future intelligence, depend not only on his diet, especially protein, from the very beginning of his existence, but also on his mother's and even his father's; perhaps even on the diet of several preceding generations. If these facts are correct, then we must consider the consequence of today's human evolution: by the year 2000, 80% of the human race will be either underfed or born of underfed parents. Referring to the soaring birth rates in the Third World, Dr Candau, Director General of WHO, declared in April, 1973: "Malnutrition of the children in the Third World can only contribute in the immediate future to mediocrity in the school children; in time, this constitutes a serious mortgage on the future".

The nutrition imbalance therefore is also very serious. It now seems inevitable that the great majority of the 3500 million additional population which will occupy the planet between now and the year 2000 will certainly be underfed and psychologically deficient.

In fact, the population growth rate in the Third World is at least 2·5%. It will therefore be necessary to achieve an annual increase in food production at the same rate if only to maintain the present ratio, which is obviously insufficient, especially in quality. All nutritionists are agreed in their demand for more food production.

Professor Alexis Pokrovsy, of the USSR Academy of Sciences, Director of the Food Institute, goes so far as to recommend four times the present output of food. This would increase the protein rate from the present 58 grams daily to 100 grams, estimated as desirable by the Soviet Union.

This fourfold increase in 30 years, which amounts to an annual food production growth of 4·5%, is literally impossible for five reasons:

1. A constant agricultural growth of more than 2% per annum (especially in the tropics), has never been known.
2. The hopes — somewhat premature — raised by the "green revolution" have not been fulfilled because the use of selected varieties of wheat and rice, necessitating large quantities of fertilisers and elaborate farming methods, requires large investments.
3. The best fields are already under cultivation; moreover, especially in Asia, there is no more new land to be cultivated. In Africa and Latin America there is, of course, virgin terrain, but it is poor in quality, difficult to cultivate and generally unsuited to green revolution requirements (wheat production is not possible, and the rice fields are very limited in extent).
4. Only educated men, trained and well equipped farmers can increase agricultural productivity. The cultural imbalance, which we shall examine, presents a very serious problem for tomorrow's humanity; and for some decades it will prevent agricultural production from satisfying the needs of all men.
5. All the new nutritional processes, such as petrol, algae, proteins from the sea, etc, are still only in the minds of the technologists or in the laboratory stage. Until now, not a single man in the world has agreed to eat synthetic foods or learned to appreciate their value. Therefore, positive results in the world's food balance cannot be expected prior to 2000.

Thus, it seems certain that during the next five decades, man will suffer from a very serious food imbalance. In the year 2000, there will be 1500 million rich, overfed people and 4500 million poor, underfed and sometimes famished souls, especially in Asia.

Education

The educational imbalance, which will beset the world for an unpredictable length of time, will give added proof every day that it is one of mankind's most chronic ailments. It is impossible today to establish with any accuracy the number of illiterate individuals in the Third World; all the figures released by UNESCO are incorrect because only the statistics turned in by its sovereign members can be registered. Yet any traveller in the Third World who estimates the illiterate at 98%, ie those who do not practise reading and writing every day, can be sure that his evaluation is fairly accurate.

Progress in this sector is difficult for four reasons:

1. In the initial colonial and post-colonial phase, the mere introduction of the developed nations' teaching methods was felt to be fully adequate. The results, however, showed it to be a total failure.
2. No one, either in UNESCO, or in the countries concerned, has yet conceived of teaching, educational or training methods (three separate sectors but complementary and indispensable) adaptable to vast numbers and categories of men, such as the Latin American, the African Black, the Arab, etc. Everything still remains to be done.
3. Modern teaching instruments, such as audiovisual equipment and communications satellites, are still only projects. Even if they were applied, there are no subjects to teach, no methods of instruction, no educational systems programmed. Nothing has been done in this area.
4. The population expansion is so great that it is not humanly possible to train the number of teachers required, build the necessary schools, etc. Evidence of this is given by the case of Morocco which, although a developed country, has been obliged to import six thousand teachers from France.

We can gain a clear idea of the worldwide harm caused by cultural imbalances if we ponder the words of Tunisia's President Bourguiba, delivered on March 19, 1973, before the Congress of Arab Writers:

The struggle between Arabs and Zionists is primarily a problem of scientific and technological inequality. It is not [Arab] culture which is weakening in its essence, but rather the underdevelopment of the [Arab] nations which is impairing all their vital forces . . .

Speaking of the developed world, he added: "Those who are creating this magic new world are all non-Arabs".

Thus one can fathom the sense of alienation felt by the peoples of the Third World, faced by the development of modern civilisation.

The rural—urban flow

Another world imbalance deserves mention even though it exists within each Third World country, unlike the other imbalances we have examined, between industrial and under developed countries: urban/rural imbalance.

The growth of cities is a general phenomenon, but it does not assume the same character in the industrial nations as it does in the countries of the Third World. In the advanced nations, the growth of agricultural productivity liberates manpower ("agriculture is not short of hands; it has too many"); the unwanted agricultural workers, therefore, move to the cities. In the Third World, the continuing flight to the cities is not caused by agricultural growth — which does not exist — but by the extreme poverty of the rural masses in constant demographic expansion. The poorest peasants, the most disinherited, prefer the misery of the city to the hardships of the fields. This is what the Mayor of Sao Paulo (which receives 400 000 poor peasants each year instead of the specialised workers it needs) recently pointed out. In Dakar, as in Casablanca and Calcutta, the influx consists of the disinherited. Moreover, those rural young who have received a smattering of education — often sketchy — feel that their place is in the city. Obviously, these urban migrations worsen the problem of food shortages.

The French Institute of Demographic Studies Bulletin writes:

It must be noted that the rapid spread of education among the peasant children in the Third World is intensifying the rural exodus, and consequently accelerating urban growth, with the result that urban unemployment is rising.

This is a very serious matter. Experts estimate that there are already dozens of millions of urban unemployed, men who never work and therefore depend upon others lucky enough to draw a scanty wage. Such a situation provokes a host of easily imaginable problems, including the "faecal peril" mentioned by Yves Lacoste, which is already threatening various metropolitan centres in Asia.

Political power

The last disparity to be mentioned is the political imbalance. This results from the political explosion which followed the last World War and spawned a number of micro-nations that emerged from the dismembered colonial regimes. The Third World thus found itself fragmented and obliged to confront the industrialised countries (Western Europe, the European socialist bloc), which were concentrating and expanding their economic, financial and military power as a consequence of their growth. Furthermore, the great powers inaugurated a new system described as aid and cooperation on behalf of the Third World. Unfortunately, this undertaking has too often assumed the form of neo-domination or neo-colonialism. The latter is sometimes commercial and economic in substance, as is especially the case in Latin America. But it also rapidly becomes political and military, as in the case of the Near East and South-East Asia.

Political imbalance is one of the most serious threats today. Not only does it imperil world peace in many parts of the planet, but it also aborts all the development policies designed to help the Third World. Obliged to negotiate its cooperation agreements separately with the powerful states, every small state invariably finds itself in a position of inferiority. When bargaining begins, its sovereignty and dignity are put in constant jeopardy.

How can these countries seriously undertake to build their nation and plan their future when every year they must beg for agreements which demand that they make considerable commercial, financial, cultural, diplomatic and often military concessions? This system of so-called bilateral cooperation between a great nation and a small nation cannot continue.

A way forward

Such are the imbalances which prevail on this planet — evident or latent, present or future, all inevitable and growing, since the very forces which have given rise to them are either expanding or stagnating. And each one of these imbalances is a potential source of tension, conflict and disorder of the spirit — such as the psycho-sociological, passion-rent situation in the Middle East — and hence a possible cause of war.

To say that the gap between the two blocs is widening is to declare an obvious fact; but this image states only in terms of the GNP what is really the sum total of all the major imbalances — far more serious than merely what is happening on the consumption level. This situation is tragic, not only because it affects millions of men and raises a formidable number of problems, but also — and especially — because it appears to concern those responsible for the active world only superficially.

The Third World problem seems to be one of a separate world, of a world "which doesn't interest us". Public opinion polls show, however, that the public is not indifferent to the scourge of hunger and suffering; yet it is extremely rare to find sympathetic men who can think not only in terms of their immediate, national and international environment, but also in terms of the planetary environment.

Statesmen think of the future on a very short-term basis and in terms of an environment which loses interest as distances lengthen. Businessmen's interests are limited to their own market and that of the few countries they deal with. As for international officials, they hold technocratic, diplomatic or simply bureaucratic opinions but their real influence is minimal. On the whole, no authority in the world is really thinking of the world problem.

With sensational originality, the Club of Rome has brought up, for the first time, the problem of the planet on the planetary scale. It has demonstrated, that if our planet is not the centre of the universe, as Copernicus showed, at least it must be the centre of our preoccupations.

This is why the Club of Rome's conclusion No. 3 on the MIT report must be recorded:

We recognise that a substantial *improvement in the lot of the nations in the process of development is the* sine qua non *condition for the new world balance.*
Here, for the first time, a scientific research project has concluded that, without the development of the Third World, world balance is impossible.
We maintain that only a strategy on the planetary scale can make it possible to realise such a programme.
Without such action, present inequalities will only sharpen dangerously. We are rushing toward disaster if, in their sections, the nations continue to pursue only their own interests or if the industrialised countries and those in the process of development enter openly into conflict.

The Club's conclusion No. 6 adds:

We are unanimously convinced that the first task which imposes itself on humanity is the rapid and radical redress of the present imbalance which is dangerously widening . . . Entirely new methods are indispensable in order to give the various societies objectives aiming at the search for successive states of balance rather than continuous growth. Such a reorganisation requires an intense effort of understanding, of imagination and the definition of a new political attitude and a new ethic. We believe that this effort can be made.
This supreme effort is the challenge offered our generation. The enterprise must be resolutely started without delay and a sharp turn must be made before the end of the next ten years . . .

The italicised sentences seem to us to be fundamental. Each word must be weighed.

The quest for a new ethic

Hundred of books and thousands of reports have been written on the problem of the Third World. None has yet clearly stated the question which common sense calls for: what can be done?

As the Club of Rome points out, new methods are indispensable, requiring an intensive effort of understanding and imagination and the definition of a new ethic.

We shall attempt to define the main overall factors of a possible ethic following three main tenets.

1. *The world should evolve towards a plurality of civilisations rather than a universal civilisation.* Until recently, the development model of the Western industrial societies was not contested; it was considered the only possible road to development.

In the industrial sector, there were two predominating ideologies, apparently mutually opposed: capitalism and socialism. But with the evolution of organisational techniques, it became clear that their methods tended to come progressively closer. The efficiency of the market economy was no longer disputed. The social justice of planned economy was more and more appreciated. Thus, cooperation between the two systems became wholly conceivable, and actually began in the seventies.

However, two-thirds of all mankind remained totally excluded from this technological civilisation and its development processes. At the end of World War II, the United Nations' top economists accurately calculated the amount of capital necessary for the Third World to embark on the process of development and allow them to catch up with the industrial nations. We can safely say that after 30 years of efforts, this concept has completely failed; meanwhile, the gap separating the two worlds has widened relentlessly.

What is more, in this decade, the industrial civilisation system has been contested by its own youth, and we cannot yet tell whether or not their confrontation movement will radically or moderately affect the growth curves projected by the futurologists until the year 2000.

What is certain is that the Western industrial civilisation, whether capitalist or socialist, cannot claim to serve as the single, inevitable model for the civilisations of the Third World, which will represent four-fifths of humanity by the year 2000.

We must abandon the idea of a universal, technological civilisation engulfing the whole world in a uniform way of life for a planetary development of a plurality of civilisations, not based exclusively on the growth of material consumption.

This is to say that by exploiting a uniform technology, we should see the development of different societies and different civilisations which would offer man true possibilities of development and expansion, in a word, a framework more attractive and stimulating than global uniformity.

Both capitalists and socialists today are asking the same question: what kind of civilisation are we seeking? What "civilisation project" must we establish as the objective of our growth policies?

For the Third World countries, which for a long time evinced no reaction to this problem and, again for a long time imitated, sheep-like, the dominant nations, the moment has come to ideate and choose their own projects of civilisation.

Stated differently, the following basic question may now be asked: in order to build the Africa of tomorrow, the South-East Asia of the year 2000 and the Arab world . . . should these regions emulate 19th-century European and American development, beginning with the steam engine, heavy industry, huge textile mills, etc, or should they rather devise other methods specifically adapted to their particular abilities as well as to modern technology?

Obviously, the second alternative is the only logical one. Hitherto, the fragmentation of the Third World into a number of micro-nations emerging from colonial domination has made it impossible to set up authentic "civilisation projects". No Third World country is big enough to conceive a civilisation. Economically, Dahomey is roughly equivalent to a small French town like Nantes; Senegal is "smaller" than any French *département*; while the whole of Africa counts for less than Italy: 2% of the world's GNP against 3·5%. Even with its population of 88 million, enormous Brazil claims an economy which is barely two-thirds that of Spain, with a population of 34 million.

Under these circumstances, how can we think of moulding great concepts for a civilisation in Black Africa, a civilisation in Latin America and a modern Arab civilisation when the frameworks allow so little room to work in? Only within the range of the great human communities can great civilisation projects arise, since they constitute the primary condition for a true policy and a true development strategy.

2. *The world consists of ten great human and natural communities.* If we accept the fundamental principle that the development of nations must be achieved within the framework of the great communities, as Europe is evolving today, then we can readily distinguish ten large, existing groups, characterised by:
- a great human homogeneity
- a sufficient area for expansion
- a large consumer market

These groups comprise four industrial communities: North America, Western Europe, the USSR and Japan and six communities of the Third World: Latin America, Black Africa, Maghreb-Near East, India, South-East Asia and China.

The extraordinary homogeneity of these six Third World communities, the unity of their vast geographical expanses and the internal complementary nature of their

41

regions are immediately evident.

Each of these Great World communities will have to define its own "civilisation project". It seems to us that the very first task for each should be to develop its project, define the stages, the means and methods to be adopted and its relationships with other communities etc in the Council of the Wise. Swamped in the sea of free exchange, no Third World nation is capable of structuring its own development economy. The USSR would never have attained its present industrial strength had it opened its doors to American, European and Japanese commodities.

Only an integrated development within a well-defined community has enabled the USSR and China to develop, as was the equally remarkable case of Europe. Indeed, the development of a country is "the development of the whole man and of all men"[4], and the development of all women and children, not of its capital city and a few industrial plants. It is the development of the whole human community.

From this point of view, we can safely conclude that development in Brazil is a failure, as it is in the Ivory Coast, in India and in the whole of the Third World. We believe with the Club of Rome that a new ethic is essential.

The creation of the Third World's great communities seems to be the primary objective to be defined, studied, organised, encouraged and developed. The Latin-American community is already considered by many to be the entire continent's primary objective. In Black Africa, two communities are beginning to emerge, West and East Africa. They urgently need encouragement. In the Maghreb-Near East, from Casablanca to Teheran, a whole world is ready to rise above several millennia of discord; it is the only solution to an absurd, endless conflict. South-East Asia will find no solution other than the formation of a large community, making reconstruction possible and a new world for generations that have known only the horrors of war. India, too, together with Pakistan, Ceylon and Bangladesh, must one day come to realise that only economic unity will settle the turbulent problems of the whole sub-continent, to the benefit of countless unhappy men, who have a right to lead a decent life on this planet.

We must point out here that the larger the community, the more it will feel internally balanced, and the more it will shrug off its imbalances in relation to the outside world. The Chinese give us an excellent example of such internal stability.

We can trust a world system based on ten great communities to provide a more peaceful life for men on earth. The example of united Europe is typical: for a long time the scene of the most violent conflicts, this continent has become, in the eyes of the world, an area where peace is most secure. An Egyptian leader recently declared that "Europe is the greatest hope of the Third World, because it is the only great force in the World that is non-nationalistic and by its very essence, non-imperialistic".

3. *Cooperation between the great communities of the world is the only way to establish a true development strategy.* From a realistic viewpoint, it is an illusion to assume that both the industrial world and the Third World are prepared to conceive and realise a world development strategy. This would require a world super-authority and a world government, which cannot reasonably be foreseen before several decades.

Moreover, on the methodological level, men are not yet confident enough that they can develop and activate a gigantic organisation which would include – under computer control – a systematic development of some 30 industrial nations on the one hand and some 100 Third World countries on the other. This is a too techno-cratic vision of a world still ridden with passions, prejudices and differences.

Nevertheless, regional cooperation policies established among the great industrial communities (which number four) and among the communities of the Third World (which number six) would appear to be much more realistic.

Certainly, no Third World development can be really possible without massive material contributions from the four industrial communities. Considering the Third World's demographic growth rate, ranging between 2·5 and 3·5%, these countries cannot, through their own efforts, ensure the acquisition of sufficient capital to subsidise an adequate development, even in the hypothetical case of a civilisation not based exclusively on material consumption.

With a 2·7% growth rate, the Third World countries must necessarily invest 12% of their GNP annually only to maintain their present standard of living. They would need to invest 21% of the GNP to achieve a 3% growth in their standard of living (which would only double the individual's very low standard of living in 25 years). These figures are evidently theoretical in the new development concept, nevertheless they provide a useful scale of measurement.

It is therefore essential that a sizeable part of the industrial countries' growth be earmarked to equip and develop the Third World. Let us say that the size of such an operation should involve not 1% of the industrial countries' GNP but rather 2% to satisfy the realistic needs of the Third World. This amounts to a considerable contribution – actually 10% of the developed countries' annual capital formation, or one-tenth of all their investments in industry, real estate, education, hospitals, etc. In plain figures, such a contribution would total approximately $50 thousand million annually.

To estimate the extent of this yearly contribution, let us say that:

public aid to the Third World now totals	$8 thousand million
public and private aid	$15 thousand million
yearly expenses for armaments	$150 thousand million

We can see, then, that an annual contribution of $50 thousand million by the industrial countries for Third World development would not be unreasonable. It would only mean tripling the current contribution by reducing present armaments expenses by a mere 20%.

The problem is that today no world structure exists to receive the $50 thousand million (rivalries among the big powers are still too strong), nor is there any organism to distribute these sums among the more than 100 recipient countries, from Bolivia to Rwanda, from Gabon to Bangladesh, etc.

It is therefore altogether evident that the $50 thousand million thus distributed – according to political or administrative criteria – would simply be wasted as a result of non-existent channels. This is why it would seem that only regional negotiations are likely to be effective in carrying out vast, well-prepared aid programmes.

Let us not delude ourselves. Public opinion in the industrialised capitalist and socialist countries would not support the idea of a contribution equivalent to 2% of the GNP — or 10% of these nations' investment capacities — without a solid guarantee that the funds would be efficiently and profitably invested.

Let us take two examples:

• 30 to 40% of European aid — $18 thousand million if we calculate the contribution at 2% of the GNP — could be assigned to Black Africa and its utilisation discussed directly by the communities. This would not exclude a coordinated contribution from the USA, Japan and possibly even from the USSR.

• Aid to the Maghreb-Near East could likewise be negotiated from community to community, with the same cooperation from the other industrialised nations.

Table 1 shows percentages of world GNP as at 1971, excluding the Maghreb-Near East, for simplification. Those countries marked with an asterisk have "super-productive" capacity, ie an available margin in industrial equipment which could be allotted to the Third World without reducing domestic supplies. Such capacity should be mobilised for a systematic programme of aid to the Third World.

Let us repeat this *sine qua non* principle: without the industrialised world's massive cooperation, it is impossible to envisage any improvement in the Third World's situation. Without that cooperation, chronic global imbalance is inevitable; and this means catastrophe for the modern world.

Our conclusion on this "quest for a new ethic" may be summed up as follows:

1. Our technical civilisation is being contested today. Certainly, therefore, it cannot serve as a model for three-quarters of mankind now or four-fifths by the year 2000.

2. It is a matter of urgency that the countries of the Third World — divided today into micro-nations — realise the extreme necessity of organising their own development, according to methods and objectives they must establish themselves. Why should the countries of the Third World join in a frantic race — most likely hopeless — to catch up with industrial civilisations, which are not even sure themselves of being on the right track?

TABLE 1. 1971 PERCENTAGES OF WORLD GNP

	%
USA*	35
Western Europe*	26
USSR*	13
Japan*	7
Latin America	5
South-East Asia	4
China	4
Africa	2
Australia*	1

* Countries with a "super-productive" capacity.

3. The framework of the six great communities of the Third World which we have defined can offer those countries an opportunity to expand an authentic development policy and help them to overcome their nationalistic mentality and realise their deep aspirations.

4. Whatever "civilisation project" each great community works out, a massive systematic contribution by the four great industrial countries ($2400 thousand million of their 1971 GNP) constitutes the only possible means of financing the equipment required for the Third World's development.

5. For obvious political reasons, this contribution — which could reach $50 thousand million a year — cannot be organised on a world scale before several decades. To hope for worldwide politics today is a utopian dream.

It is only through agreements between the industrial and Third World communities that a realistic, enduring understanding, essential to the realisation of long-term development projects, may be achieved.

In this respect, an agreement between the European Economic Community and the West African Community, now under discussion[5], might soon serve as a model for such a development strategy, which is being much debated, although it still remains to be defined.

6. The principle of communities would seem to be the basic cornerstone of the civilisation projects we have mentioned.

On this principle, it will be possible to lay the foundation of an organised world which is diversified, more human and more responsive to the aspirations of all peoples, rather than a unified world subjected to the technocratic laws of the consumer societies.

Redressing the balance

In conclusion, the following two questions must be posed: will the new ethic we have summarily outlined make it possible to reduce the imbalances described above and will this project for a new ethic help orient the world towards a better life?

We shall try to answer point by point on the five principal imbalances.

Population

Within the framework of the proposed ethic, each community itself must calculate the impact of its demographic growth rate on its development project.

The Chinese, who constitute a community already organised, fully understand this problem; hence they have assumed the responsibility for their own development. They have understood that, by calling for fewer births — via methods they have devised themselves — they could step up the rhythm of development. The small countries, instead, are indifferent to this problem — they are naively proud of their large populations — because they lack a sense of global responsibility for their development. They are more concerned with their immediate problems, such as the market price of coffee or groundnuts, rather the problems of tomorrow.

We have also seen that the incentive to reduce the birth rate apparently comes only with a rising standard of living especially with a higher cultural level, which enlightens parents on the seriousness of their responsibilities. But we have also seen that development cannot be expected with 2·5–3·5% demographic growth rates. What can be done?

Within the framework of the great communities which we have proposed, it will behove the responsible authorities to organise village rural life according to the terms of the "village communities" formula, which requires only small investments to raise the standard of living in the villages, improve the education of children and the training of their parents and organise life in the villages; in other words, to make the village a cultural centre in the broadest sense of the word.

In a recent inquiry conducted in India, where such village communities were established by Pandit Nehru, the following question was asked[6]: How many children per family are there in the village? The educated answered three or four, the illiterate five or six. Ten years ago they would have said six or seven.

Village communities have been expanding rapidly, thanks to education, hygiene and agriculture monitors – three technicians, one educator, one medical worker and one agronomist can assist 100 to 200 inhabitants. This rapid expansion must exert an influence on the cultural and city/country imbalances, the gravity of which we have pointed out.

Food

This imbalance, which threatens not only famine but also widespread malnutrition together with dire psychological consequences, is the result of two factors: demographic growth, which we have just discussed, and insufficient agricultural productivity.

In the Third World, one is always surprised to hear responsible officials stress the need for agricultural progress while seeming to take no interest in the means of realising it. This is because agricultural problems in the tropics are dimly understood, and no small country feels capable of assuming responsibility for them.

One of the first concerns of the great communities will be to spur research in all the community's territory: agronomic research, practical applications of agronomy, cultural and sociological research, etc. It is not generally known that Black Africa claims not a single research centre set up to introduce and select vegetables and fowl suitable to the varying climates. Yet the Africans are in dire need of these products. It is only through a full awareness of Black Africa's total requirements that the Black African community will confront its responsibilities and undertake to include all its needs in its civilisation project.

As for introducing into the rural areas all the technical advances achieved, again it is only through the intermediary of the village communities that they can possibly be applied, as the examples of China and Israel amply demonstrate.

Education

The executive organ of each community will fulfil the task of investigating educational systems as well as adult instruction and training to comply wholly with the children's needs. No one has ever attempted this approach, not even UNESCO.

Today, each country is too small; it does not consider itself responsible for the teaching system it imported lock, stock and barrel from abroad.

Imagine the progress that the Arab world would make if all the children were educated and trained by "modern" methods, conceived and programmed according to the attributes and needs of the Arab realm today!

Also in this domain, the diffusion of education, teaching, and culture by the village communities would enable them to overcome this cultural imbalance, which is the scourge of the Third World. In his most recent book[7], René Dumont describes it aptly:

The school would seek the collective promotion of the village, with children, young people and adults united in a system of permanent education and mutual instruction . . .

Political power

Obviously, the first consequence deriving from this policy of agreements between Communities would automatically put an end to what is generally called neo-colonialism, that is, the system of aid granted by a dominating state to a dominated state.

This aid would truly become a policy of negotiated cooperation, encompassing the whole Community, its general equipment and its human investments expressed in long-term programmes.

Military aid would then be meaningless, since it can only be conceived on a state-to-state basis. It is hard to imagine the European Economic Community agreeing to finance the Maghreb-Near East Community while the USSR would be content with furnishing arms to Egypt.

The great communities system will force the states to renounce military aid, as well as diplomatic, economic and trading demands. It will put an end to the blights of nationalism.

Standard of living

With the concept of "civilisation projects" particular to each great community, appraising standards of living in terms of the GNP no longer has an absolute sense. To the commodities produced and consumed, we should have to add all the interchanged internal services, which cannot be evaluated in terms of money.

To introduce water filters in a country where the water is polluted, to hire expert agricultural advisers, introduce selected seeds of unknown vegetables — such as carrots, rich in vitamin A — in countries without vitamins for eight months out of the year, to accustom men to use fertilisers, educate children in the remotest villages and balance the diet of pregnant women and children, etc — all these things have no price and cannot figure in any GNP accounting.

Is it really the ideal objective of the Indian, the Afghan, the Madagascar or the Peruvian to consume constantly more and more, like the American, the Swede or the Parisian?

No great changes would be necessary to give proof that the life of a Moroccan, Iranian or Mexican peasant could be, after all, just as humanly satisfying as that of a labourer in Detroit or the Ruhr. Indeed, we believe that the differences, on the human scale, between under-developed and developed countries are not as striking in the quality of life as we are led to believe by the disparities shown in GNP per inhabitant statistics.

In other words, we believe that the Black Africans and the Arabs, for example, could be just as contented with a standard of living possible on $500 — to use GNP terms — provided that they are living in an adapted civilisation, as Americans with an income of $4000.

Although this example represents two different life styles in terms of quantity, surely they could be equivalent in terms of human quality?

Therefore, the undertaking we suggest should be carried out on a worldwide scale is not so very enormous. Nor is it as utopian as the proposed project which would have enabled every Indian to enjoy the same per capita standard of living as the European.

It is within our reach, depending on our will.

References

1. A. Sauvy, *Croissance zéro*, Calmann-Levy, 1973, page 114
2. P. Bairoch, *Le Tiers Monde dans l'Impasse*, Gallimard, 1971, page 260
3. It is easy to see the difficulties which are met by countries like France and Italy, confronted by a demographic growth rate of a mere 1%
4. According to the Rev Lebret, leader of the "Humanisme et Developpement" group.
5. President Leopold-Sedar Senghor is actively promoting both the West African Community, from Zaire to Mauritania, and a new Euro-African political entente
6. *Le Figaro*, January 27, 1973
7. *L'Utopie ou la Mort*, Editions du Seuil, 1973

Technics and human culture

Lewis Mumford

At an early stage in my studies on the relationship between technics and social change, I found that I could not understand many contemporary institutions and activities without tracing them back to what often proved very remote, sometimes prehistoric, events. Unfortunately most of our thinking today in technocratic circles is being done by one-generation minds bedazzled over our immediate successes with nuclear energy, moon-shots, and computers — however isolated these feats are from the total historic culture that made them possible, and from man's many non-technological needs, projects, and aspirations which give meaning and value to the whole process.

Naturally those who think this way do not thank me for pointing out that their so-called Industrial Revolution did not begin in the eighteenth century; that on the contrary the "new wave" in Western technology began as far back as the eleventh century; and that the invention of the mechanical clock in the fourteenth century did more to advance modern technics than the steam engine or the automatic loom. For the clock, on its very face, unified our whole conception of time, space, and motion, and laid the foundation in its exact measurements by standard units for the astronomical-mechanical world picture that still dominates our minds and our daily activities.

Despite later reinforcements from historians like Bertrand Gille, Georges Friedmann and Fernand Braudel, this well-documented revision of the standard technological clichés has not been generally accepted. So it would be presumptuous on my part to look for a more favourable immediate reception for the even more radical revisions I shall proceed to summarise. For I propose to lengthen the historical perspective sufficiently to present a more adequate picture of the relationship between technological progress and social change and human development. Instead of celebrating the further expansion and acceleration of technology, on the lines which have ultimately led to the power system that now governs our lives, I shall endeavour to restore our ecological and cultural equilibrium. And so, far from taking the conquest of nature and the elimination of any recognisable form of man as the inevitable consequence of technological progress, I question both the value of this one-sided conquest itself and its inevitability. Not least, I seek to expose the irrational factors that have led modern man to forfeit those essential expressions of human creativity that do not conform to the unlimited quantitative requirements and pecuniary ambitions of the power complex.

The nature of man

My point of departure in analysing technology and social change concerns the nature of man. And to begin with I reject the lingering anthropological notion, first suggested by Benjamin Franklin and Thomas Carlyle, that man can be identified, mainly if not solely, as a tool-using or tool-making animal: *homo faber.* Even Henri Bergson, a philosopher whose insights into organic change I respect, so described him. Of course man is a tool-making, utensil-shaping, machine-fabricating, environment-prospecting, technologically ingenious animal — at least that! But man is also — and quite as fundamentally — a dream-haunted, ritual-enacting, symbol-creating, speech-uttering, language-elaborating, self-organising, institution-conserving, myth-driven, love-making, god-seeking being, and his technical achievements would have remained stunted if all these other autonomous attributes had not been highly developed. Man himself, not his extraneous technological facilities, is the central fact. Contrary to Mesopotamian legend, the gods did not invent man simply to take over the unwelcome load of disagreeable labour.

All man's technological inventions are embedded in the human organism, from automation to cybernation; automatic systems, indeed, so far from being a modern discovery, are perhaps the oldest of nature's devices, for the automatic responses of the hormones, the endocrines, and the reflexes antedated by millions of years that super-computer we call the forebrain, or neopallium. Yet anything that can be called human culture has demanded certain specific technical traits: specialisation, standardisation, repetitive practice; and it was early man's positive enjoyment of repetition, a trait still shown by very young children, as every parent knows, that underlay every other cultural invention, above all language. It was this studious technical development of the organism as a whole, not just the employment of man's hands as facile tools or tool-shapers, that accounts for the extraordinary advances of *homo sapiens.* In making these first technological innovations man made no attempt to modify his environment, still less to conquer nature: for the only environment over which he could exercise effective command, without extraneous tools, was that which lay nearest him: his own body, operating under the direction of his highly activated brain, busy by night in dreams as well as by day in seeking food or finding shelter.

On this reading, before man could take even the first timid steps toward "conquering nature", he first had the job of controlling and utilising more effectively his own organic capacities. By his studious exploration and reconstruction of his bodily functions he opened up a wide range of possibilities not programmed, as with other animals, in his genes. Strangely, it took André Varagnac, a French researcher of the archaic folk remains of neolithic culture, to point out that the earliest mode of a specifically human technology was almost certainly the technology of the body. This consisted in the deliberate remodelling of man's organs by enlarging their capacity for symbolic expression and communal intercourse. Most significantly, the only organ that greatly increased in size and weight was the brain. By this close attention to his body, even primitive man at a very early moment placed his automatic functions under some measure of cerebral intervention: the first step in conscious self-organisation, rational direction, and moral control. Man is his own supreme artifact.

Long before man had given stone tools even the crudest form of a hand-axe, he had already achieved an advanced technology of the body. These basic technical achievements started with infant training; and they involved not only repetition, but foresight, feedback, and attentive learning: learning to walk, learning to control the excretory functions, learning to make standardised gestures and sounds, whose recognition and remembrance by other members of the group gave continuity to the whole human tradition. Not least man learned to distinguish to some degree between his private often subconscious dreams and his shared waking realities; and as the forebrain exercised more authority he learned, likewise in the interest of group survival, to curb his blind destructive impulses, to diminish overpowering rage and fear, to inhibit demented fantasies and murderous aggressions, and to superimpose social order and moral sensibility upon random sexuality. Unless man had mastered fear sufficiently to be able to play with fire — a feat no other animal has dared to perform — he would have lacked one of the essential requirements for the survival and spread of his species, since fire enormously increased the number of foods and habitats available for both paleolithic and neolithic man.

The point I am stressing here is that every form of technics has its seat in the human organism; and without man's many artful subjective contributions, the brute energies of the existing physical world would have contributed nothing whatever to technology. So it was through the general culture of the human body, not just through tool-making and tool-using, that not only man's intelligence but other equally valuable capacities developed. This culture, with its symbolic expressions, its moral selectivity and direction, its emotional communion — in short, its life wisdom — was more essential to human development than any special technical advances from ancient fire-making and tool-making to modern automation, mass production, and cybernation. Almost down to the present century, all technical operations took place within this organic and human matrix. Only the most degraded forms of work, like mining, which was reserved for slaves and deliberately treated as punishment, lacked these happy human qualities. Today at last we are beginning to measure the loss we face through our present efforts to remodel the human organism and the human community to conform to the external controls imposed by the power system.

This leads me to my second departure from technocratic orthodoxy. How is it that modern man since the seventeenth century has made technology the focal centre of his life? Why has the "Pentagon of Power", motivated by the conception of constant technological progress and endless pecuniary gain, taken command of every human activity? At what point did the belief in such technological progress, as a good in itself, replace all other conceptions of a desirable human destiny? To answer this question I have had to trace this power-fed aberration back five thousand years to its point of origin in the pyramid age. But first I would call attention to its modern expression in a sign that once greeted the visitor at the entrance of a World's Fair celebrating "A Century of Progress". That sign said "Science discovers: Technology executes: Man conforms".

Man conforms indeed! Where did that strange categorical imperative come from? How is it that man, who never in his personal development conformed submissively to the conditions laid down by nature, now feels obliged at the height of his powers

51

to surrender unconditionally to his own technology? I do not question the fact itself. During the last two centuries a power-centred technics has taken command of one activity after another. By now a large sector of the population of the planet feels uneasy, deprived and neglected — indeed cut off from "reality" — unless it is securely attached to some part of the megamachine: to an assembly line, a conveyor belt, a motor car, a radio or a television station, a computer, or a space capsule. To confirm this attachment and make universal this dependence, every autonomous activity, once located mainly in the human organism or in the social group, has either been wiped out of existence or reshaped by training and indoctrination and corporate organisation to conform to the requirements of the machine. Is it not strange that our technocratic masters recognise no significant life processes or human ends except those that further the expansion of their authority and their magical prerogatives?

Thus the condition of man today, I have suggested in *The Pentagon of Power*, resembles the pathetic state of Dr Bruno Bettelheim's psychiatric patient: a little boy of nine who conceived that he was run by machines. "So controlling was this belief", Dr Bettelheim reports, that the pathetic child "carried with him an elaborate life-support system made up of radio, tubes, light bulbs, and a breathing machine. At meals he ran imaginary wires from a wall socket to himself, so his food could be digested. His bed was rigged with batteries, a loud speaker, and other improvised equipment to keep him alive while he slept".

The fantasy of this autistic little boy is the state that modern man is fast approaching in actual life, without as yet realising how pathological it is to be cut off from his own innate resources for living, and to feel no reassuring tie with the natural world or his own fellows unless he is connected to the power system, or with some actual machine, constantly receiving information, direction, stimulation, and sedation from a central external source, with only a minimal opportunity for self-motivated and self-directed activity.

Technocratic man is no longer at home with life, or with the environment of life; which means that he is no longer at home with himself. He has become, to paraphrase A. E. Housman, "a stranger and afraid" in a world his own technology has made. But in view of the fact that during the last century our insight into the organic world has been immensely deepened, indeed revolutionised by the biological sciences, why do we still take the Newtonian "machine" instead of the Darwinian "organism" as our model, and pay more respect to the computer than to the immense historical store of knowledge and culture that made its invention possible?

Since my own analysis of technology begins, not with the physical phenomena of mass and motion, but with organisms, living societies, and human reactions, I do not regard such conformity as anything more than an institutionalised mental derangement, one of many errors that the human race has committed while straining to improve its condition and to make use of powers and functions it does not even now fully understand. Within the framework of history and ecology one discovers a quite different picture of Nature, and a more hopeful view of man's own dynamic potentialities. Biology teaches us that man is part of an immense cosmic and ecological complex, in which power alone, whether exhibited as energy or productivity or human control, plays necessarily a subordinate and sometimes inimical part, as in tornadoes and earthquakes. This organic complex is indescribably rich, varied, many-dimensioned, self-activating; for every organism, by its very nature, is the focal point of autonomous changes and transformations that began in the distant past and will outlive the narrow lifespan of any individual group, or culture. What is now accepted and even exalted as "instant culture" – the beliefs and practices of a single generation – is in reality a blackout of collective memory, similar to what takes place under certain drugs. This bears no resemblance at all to any human culture, since without some of paleolithic man's basic inventions, above all language, even the latest scientific discoveries of this one-generation culture could not be kept in mind long enough to be described, understood, or continued beyond their own ephemeral lifetime.

On this interpretation the most important goal for technology is not to extend further the province of the machine, not to accelerate the transformation of scientific discoveries into profit-making inventions, not to increase the output of kaleidoscopic technological novelties and dictatorial fashions; not to put all human activities under the surveillance and control of the computer – in short, not to rivet together the still separate parts of the planetary megamachine, so that there will be no possibility of escaping it. No: the important goal for all human agencies today, and not least for technology itself, is how to bring back the autonomous attributes of life to a culture that, without them, will not be able to survive the destructive and irrational forces that its original technical achievements generated. If our main problem today turns out to be that of controlling technological irrationalism, it should be obvious that no answer can come from technology. The old Roman question – Who shall control the controller? – has now come back to us in a new and more difficult form. For what if the controllers, too, have become irrational?

Discovering the "megamachine"

What then was the origin of the Victorian notion that science and technology, if sufficiently developed, would replace or happily demolish all the earlier phases of human culture? Why did "progressive" but still "human" minds, from the eighteenth century on, think that it was possible and desirable – indeed imperative – to wipe out every trace of the past, and thus to replace an organic culture, full of active ingredients derived from many ancient natural and human sources, by an up-to-date manufactured substitute, devoid of aesthetic, ethical, or religious values, or indeed any specific human qualities except those that served the machine? By the middle of the nineteenth century this belief had become a commonplace. Progress meant, not humanisation, as in the earliest technology of the body, but mechanisation; with bodily efforts becoming more and more superfluous until they might either be eliminated, or at best transferred, in a limited way, to sport and play. Was this the inevitable effect of the Industrial Revolution? And, if so, what made progressive minds embrace so fatalistically the "inevitability of the inevitable"?

My own generation, I confess, still accepted readily – all too readily – this faith in the redemptive power of science and technology: though not, I hasten to add, with quite the fanatical devoutness of a Buckminster Fuller or a Marshall McLuhan today. So, when I wrote *Technics and Civilization* more than thirty years ago, I still properly stressed the more beneficent motives and the more sanguine contributions of modern technology; and though I gave due attention to the ecological depredations of the paleotechnic phases, I supposed that these life-vitiating practices would be wiped out by the further neotechnical improvements promised by hydroelectric power, scientific planning, industrial decentralisation, and the regional city. Still, even in *Technics and Civilization*, I devoted a long chapter to the negative components which, so far from disappearing, were already becoming more demonic, more threatening, and more insistent.

Some twenty years later, in a seminar I conducted at the Massachusetts Institute of Technology, I critically reviewed this early interpretation, and found that the chapter I had devoted to the negative aspects of modern technology would have to be expanded. While all the current praise of industrial rationalisation was, up to a point, sound, this had been accompanied by an irrational factor we had not dared to face — or had mistakenly attributed to fascism or communism alone. For during the last half-century, all the concealed symptoms of irrational behaviour had suddenly exploded in our faces. This period witnessed not only the unparalleled destruction wrought by two global wars, but the further degeneration of war itself into deliberate genocide, directed not against armies but against the entire population of the enemy country. Within a single generation, less than thirty years, thanks to purely technological advances, from the airplane to napalm and nuclear bombs, all the moral safeguards mankind had erected against random extermination had been broken down. If this was technology's boasted conquest of nature, the chief victim of that conquest, it turned out, was man himself.

With these massive miscarriages of civilisation in view, I tentatively put to myself a decade ago a question that I did not ask publicly until I wrote the first volume of *The Myth of the Machine*: "Is the association of inordinate power and productivity with equally inordinate hostility, violence, and destructiveness, a purely accidental one?" This question was so uncomfortable to entertain, so contrary to the complacent expectations of our technocratic culture, that I cannot pretend that I eagerly searched about for an answer. But fortunately, at that moment I was making an intensive study of the whole process or urbanisation, that which Gordon Childe called the Urban Revolution, as it took place in Egypt and Mesopotamia toward the end of the Fifth Millenium before the Christian era. Digging mentally around those urban ruins, I discovered an extraordinary complex machine which turned out, on analysis, to be the first real machine, and the archetype of all later machines. This artifact had for long remained invisible, because it was composed entirely of highly specialised and mechanised human parts. Only the massive results of its operation remained visible, not the formative ideas and mythical projections that had brought this machine into existence.

What Childe called the Urban Revolution was only an incident in the assemblage of the "megamachine", as I chose to call it. Please note that the superb technological achievements of this gigantic machine owed nothing, at the beginning, to any ordinary mechanical invention: some of its greatest structures, the pyramids of Egypt, were erected without even the aid of a wheeled wagon or a pulley or a derrick. What brought the megamachine into existence was not an ordinary invention but an awesome expansion of the human mind in many different areas: a transformation comparable to that which took place when in the distant past the structure of language had advanced sufficiently to identify and communicate and pass on to later generations every part of a community's experience.

The decisive tools that made this machine possible were likewise new inventions of the mind: astronomical observation and mathematical notation, the art of the carved and the written record, the religious concept of a universal order derived from close observation of the heavens and giving authority — the authority of the gods — to a single commanding figure, the king, he who had once been merely a hunting chief. At this point the notion of an absolute cosmic order coalesced with the idea of a human order which shared in its god-like attributes. Then both the machine and the Myth of the Machine were born. And therewith, large populations hitherto isolated and scattered could be organised and put to work, on a scale never before conceivable, with a technical adroitness and a machine-like precision and perfection never before possible. Small wonder that those divine powers were worshipped and their absolute rulers obeyed!

In unearthing this invisible megamachine I was not so much trespassing on the diggings of established archaeologists as flying over them. So far I was safe! But my next move, in equating the ancient megamachine with the technological complex of our own time, caused me to push into heavily-defended territory, where few competent colleagues have as yet been willing to venture. This is not the place to summarise all the evidence I have marshalled in *The City in History* and in *The Pentagon of Power*. Enough to point out that the original institutional components of the Pentagon of Power are still with us, operating more relentlessly if not more efficiently than ever before: the army, the bureaucracy, the engineering corps, the scientific élite — once called priests, magicians, and soothsayers — and, not least, the ultimate Decision Maker, The Divine King, today called the Dictator, the Chief of Staff, the Party Secretary, or the President: tomorrow the Omnicomputer.

Once I had discovered the megamachine, I had for the first time a clue to many of the irrational factors that have undermined every civilisation and that now threaten, on a scale inconceivable before, to destroy the ecological balance of the whole planet. For from the beginning, it was plain, the Invisible Machine had taken two contrasting forms, that of the labour machine and that of the war machine: the first potentially constructive and life-supporting, the other destructive, savagely life-negating. Both machines were products of the same original myth, which gave to a purely human organisation and an all-too-human ruler an absolute authority derived from the cosmos itself. To revolt against that system, to question its moral authority, or to try to withdraw from it, was disobedience to the gods. Under very thin disguises, those gods are still with us. And their commands are more irresistible than ever before.

Since the original labour machine could not be economically put to work except for large-scale operations, smaller, more serviceable and manageable machines of wood and brass and iron were in time invented as useful auxiliaries to the Invisible Machine. But the archetype itself persisted in its negative, military form: the army and the army's "table of organisation" was transmitted through history, more or less intact, from one large territorial organisation to another — the army with its hierarchical chain of command, with its system of remote control, with its regimentation of human responses, ensuring absolute obedience to the word of command, with its readiness to impose punishment and inflict death to ensure conformity to the Sovereign Power. Not only does this power system break down human resistance and deliberately extirpate the communal institutions that stand in its way, but it seeks to extend'both its political rule and its territorial boundaries; for power, whether technological, political, or pecuniary, recognises no necessary organic limits.

The real gains in law, order, craftsmanship, and economic productivity the megamachine made possible must not be belittled. But unfortunately these gains were reduced, often entirely cancelled out, by the brutalising and dehumanising institutions that the military megamachine brought into existence: organised war, slavery, class expropriation and exploitation, and extensive collective extermination. In terms of human development, these evil institutions have no rational foundation or humane justification. This, I take it, is the basic trauma of civilisation itself; and the evidence for it rests on much sounder foundations than Freud's quaint concept of a mythical act of patricide. What is worse, the hallucinations of absolute power, instead of being liquidated in our time through the advance of objective scientific knowledge and democratic participation, have become more obsessive. In raising the ceiling of civilisation's constructive achievements, the megamachine likewise lowered its depths.

Technological exhibitionism

The parallels between the ancient and the modern megamachine extend even to their fantasies: in fact, it is their fantasies that must first be liquidated by rational exposure if the megamachine is to be replaced by superior and more human types of organisation and association based on personal autonomy and mutual aid. In the religious legends of the early Bronze Age, one discovers, if one reads attentively, the same irrational residue one finds in our present power system: its obsession with speed and quantitative achievements, its technological exhibitionism, its bureaucratic rigidity in organisation, its relentless military coercions and conscriptions, its hostility to autonomous processes not yet under control by a centralised authority. The subjective connection between the ancient and the modern megamachine is clear.

All the boasted inventions of our modern technology first erupted in audacious Bronze Age dreams as attributes of the gods or their earthly representatives; remote control, human flight, supersonic locomotion, instantaneous communication, automatic servo-mechanisms, germ warfare, and the wholesale extermination of large urban populations by fire and brimstone, if not nuclear fission. If you are not familiar with the religious literature of Egypt and Babylonia, you will find sufficient data in the Old Testament of the Bible to testify to the original paranoia of the power complex in the dreams and daily acts of the gods and the kings who represented that power on earth.

Just as today, unrestrained technological exhibitionism served as proof of the absolute power of the monarch and his military-bureaucratic-scientific élite. None of our present technological achievements would have surprised any earlier totalitarian rulers. Kublai Khan, who called himself Emperor of the World, boasted to Marco Polo of the automatic conveyor that brought food to his table, and of the ability of his magicians to control the weather. What our scientifically-oriented technologies have done is to make even more fabulous dreams of absolute control not only credible but probable; and in that very act it' has magnified their irrationality — that is, their divorce from ecological conditions and historical human traditions under which life of every kind, and above all conscious human life, has actually flourished. The fact that most of these ancient fantasies have turned into workaday realities does not make their present and prospective misuse less irrational.

Do not be deceived by the bright scientific label on the package. Ideologically the modern power complex, if measured by the standards of ecology and humane morality, is as obsolete as its ancient predecessor. Our present technocratic economy, for all its separate inventions, lacks the necessary dimensions of a life economy, and this is one of the reasons that the evidences of its breakdown are now becoming frighteningly visible. We have abundant biological evidence to demonstrate that life could not have survived or developed on this planet if command of physical energy alone had been the criterion of biological success. In all organic processes quality is as important as quantity, and too much is as fatal to life as too little. No species can exist without the constant aid and sustenance of thousands of other living organisms, each conforming to its own life-pattern, going through its appointed cycle of birth, growth, decay, and death. If a feeble, unarmed, vulnerable creature like man has become lord of creation, it is because he was able deliberately to mobilise all his personal capacities, including his gifts of sympathy, group loyalty, love, and parental devotion. These gifts ensured the time and attention necessary to develop his mind and pass on his specifically human traditions to his offspring.

For remember: man is not born human. What has separated man's career from that of all other species is that he needs a whole lifetime to explore and to utilise — and in rare moments to transcend — his human potentialities. When man fails to develop the arts and disciplines that bring out these human capacities, his civilised self sinks, as Giambattista Vico long ago pointed out, to a far lower level than any other animal. Since the megamachine from the beginning attached as much value to its negative components — to success in war, destruction, and extermination — as to life-promoting functions, it widened the empire of absurdity and irrationality. To face this built-in irrationality of both ancient and modern megamachines is the first step toward controlling the insensate dynamism of modern technology.

Let me cite a classic example of our present demoralising conformities. Observe what a distinguished mathematician, the late John von Neumann, said about our current addiction to scientific and technological innovations. "Technological possibilities," von Neumann said, "are irresistible to man. If he can go to the moon, he will. If he can control the climate, he will". Though von Neumann expressed some alarm over this situation, I am even more alarmed at what he took for granted. For the notion that technological possibilities are irresistible is far from obvious. On the contrary, it is a historical fact that this compulsion, except in the form imposed by the original Bronze Age model megamachine itself, is limited to modern Western man. Until now, human development was curtailed severely both by archaic institutional fixations and backward technological practices, conditioned by magical hocus-pocus. One of the chief weaknesses of traditional village communities was rather that they too stubbornly resisted even the most modest technical improvements, preferring stability and continuity to rapid change, random novelties, and possible disruption. As late as the seventeenth century an inventor in Rostock was publicly executed for designing an automatic loom.

What von Neumann was talking about was not historical man in general, but modern Western Man, Bureaucratic Man, Organisation Man, Post-historical or Anti-historical Man: in short, our compulsive, power-obsessed, machine-conditioned, contemporaries. Let us not overlook the fact that when any single impulse becomes irresistible, without regard to past experience, present needs, or future consequences, we are facing an ominously pathological derangement. If von Neumann's dictum were true, the human race is already doomed, for the governments of both the United States and Russia have been insane enough to produce nuclear weapons in quantities sufficient to exterminate mankind five times over. Is it not obvious that from the outset there has been a screw loose in the mighty megamachine? And have these paranoid obsessions not increased in direct proportion to the amount of political and physical power the system now has placed in the hands of its leaders?

How is it then, you may ask, that earlier civilisations were not destroyed by the persistent aberrations of the power complex? The most obvious answer is that their destruction repeatedly did take place, in most of the twenty-odd civilisations that Arnold Toynbee's *Study of History* examined. But in so far as these power systems survived, it was probably because they were still held back by various organic limitations: mainly because their energy, in the form of manpower, was until recently derived solely from food crops; and though sadistic emperors might massacre the populations of whole cities, this killing could be done only by hand. Even in its palmiest days the megamachine depended upon the self-maintenance of man's small, scattered, loosely organised farming villages and feudal estates, whose members were still autonomous enough to carry on even when the ruling dynasties were destroyed and their great cities were reduced to rubble.

Furthermore, between the power technics of the megamachine and the earlier organic, fertility technics of farm and garden, there fortunately persisted a third mediating mode of technics, common to both the urban and the rural environments; namely, the cumulative polytechnics of the handicrafts — pottery making, spinning, weaving, stone-carving, building, gardening, farming, animal breeding — each a rich repository of well-tested knowledge and practical experience. Whenever the centrally-controlled megamachine broke down or was defeated in war, its scattered members could re-form themselves, falling back on smaller communal and regional units, each transmitting the essential traditions of work and aesthetic mastery and moral responsibility. Not all the technical eggs were in one basket. Unitl now, this wide dispersal of working power, political intelligence, and craft experience happily mitigated the human disabilities of a system based on the abstractions of power alone.

But note: our modern power system has annihilated these safeguards, and thereby, incidentally, endangered its own existence. Thanks to its overwhelming successes in both material and intellectual productivity, the organic factors that made for ecological, technological, and human balance have been progressively reduced, and may soon be wiped out. Even as late as 1940, as the French geographer, Max Sorre pointed out, four-fifths of the population of the planet still lived in rural areas closer in their economy and way of life to a neolithic village than to a modern megalopolis. That rural factor of safety is fast vanishing, and except in backward or underdeveloped countries, has almost disappeared. No competent engineer would design a bridge with as small a factor of safety as that under which the present power system operates. The more completely automated the whole system becomes and the more extensive its centralised mode of communication and control, the narrower that margin becomes; for as the system itself becomes more completely integrated, the human components become correspondingly disintegrated and paralysed, unable to take over the functions and activities they have too submissively surrendered to the megamachine.

Judged by any rational criteria, the modern megamachine has poor chances of survival. Though everyone is now aware of its mounting series of slowdowns and breakdowns, its brownouts and blackouts, its depressions and inflations, these failures are ironically the results of the power system's very success in achieving high levels of production. Technologically speaking, the old problem of scarcity of food or goods has been solved, but the new problem of over-abundance has proved even more disconcerting, and harder to remedy without radically revising all the sacred principles of our pecuniary-pleasure economy.

The ancient megamachine worked, we now perceive, only because its benefits were reserved for a restricted, privileged class, or a small urban population. The modern megamachine, in order to justify its mass production, now seeks to impose mass production and mass consumption upon the backward populations of the planet. But it should now be plain that without deliberate human intervention, vigilantly imposing thrift, moderation, humane restraint upon the whole business of production, this affluent society is doomed to choke to death on its waste products.

The system challenged

Fortunately, in recent years there has been a sudden, if belated, awakening to the dire external consequences of our unquestioning devotion to technology's expansions and extensions. Who can now remain blind to our polluted oceans and rivers, our smog-choked air, our mountainous rubbish heaps, our sprawling automobile cemeteries, our sterilised and blasted landscapes, where the strip miner, the bulldozer, the pesticides, and the herbicides have all left their mark; the widening deserts of concrete, in motor roads and car parks, whose substitution of ceaseless locomotion for urban decentralisation daily wastes countless man-years of life in needless transportation; not least our congested, dehumanised cities where health is vitiated and depleted by the sterile daily routine. With the spread of biological knowledge that has gone on during the last generation, the meaning of all these ecological assaults has at last sunk in and begun to cause a general reversal of attitude toward the entire technological process, most markedly among the young. The claims of megatechnologies are no longer unchallengeable; their demands no longer seem irresistible. Only backward Victorian minds now believe "you can't stop progress", or that one must accept the latest devices of technology solely because they promise greater financial gains or greater national prestige.

Though there has been a general awakening to the negative goods — or "bads" as Bertrand de Jouvenel calls them — that have accompanied the explosive technology of the twentieth century, most of our contemporaries still hold stubbo.nly to the naive belief that there is a purely technological solution to every human problem. Hence the elaborate build-up, since 1945, of rocket projectiles to intercept nuclear weapons at a distance, as if this could promise any substantial control over the power-conditioned minds that had, in the first instance, produced these weapons. Such minds are open to the same kind of anti-social psychotic impulses we find spreading in many other groups. If we seriously mean to control the megamachine, we must now reverse the process that brought about its original invention, and bring back to all its human agents — not merely its leaders — the necessary self-confidence and moral discipline that will make them ready to intervene at any point where the power complex threatens human autonomy; to challenge its purposes, to reduce its automatic compulsions, to restore and further cultivate the missing components of the human personality.

This is not an empty counsel of perfection. We do not have to effect a general revolution in order to partly dismantle the megamachine and redistribute all its marvellous technical potentialities to smaller units that can exercise control. Our leaders have only to lose their religious faith in the megamachine, and return to a more organic model — a model that draws, not on one-generation experience, but on the entire technical and cultural heritage of man throughout history. Witness the exemplary decision of the American Congress to waste no further public funds on supersonic transport. This was a victory for human reason and human values over the naive ambitions of our pecuniary-power economy. That sensible decision did not need the support of a computer to sanctify its authority.

A more organic life-pattern has begun to take possession of our minds, and is laying the foundations for a conception of technology that will do justice to all the dimensions of life, past, present, and possible. This new organic model does not reject any particular mechanical organisation, but it takes the organism, or rather societies and associations of organisms and human personalities in their vast historical diversity, as the central condition of all human development. There are no artificial, prefabricated, mass-produced substitutes for life itself.

There is not space to catalogue even briefly the many separate efforts to bring our automated technology under rational criticism and effective human control. So let me cite only a single example: a fresh line of thought in economics tentatively mapped out by economists like R. H. Tawney, J. K. Galbraith, and Ezra Mishan, which challenges the notion of unlimited economic expansion. This idea actually goes back to John Stuart Mill. Instead of accepting the explosion of population, the multiplication of inventions, and the continued coalescence of corporate organisations and cartels into ever-expanding conglomerates, the new economics seeks to maintain a dynamic equilibrium to curb, or even in any particular instance to reverse, these seemingly uncontrollable processes. Whatever surplus of energies we now command would be directed, once basic human wants were satisfied, to the production of non-consumable goods and the extension of the many necessary voluntary services which cannot be performed under the canons of mass production.

The conception of such a balanced economy was put forward by Mill in his chapter on The Stationary State in his *Principles of Economics.* This chapter until now was remembered only because it was there that Mill observed that it was doubtful if all our labour-saving machinery had yet lightened the burden of the day's work for a single human being. That observation, we know, no longer holds. But Mill's forgotten suggestion that the dynamic processes of industrialisation could with human benefit be slowed down, and that economic effort should be shifted to durable products and non-consumable goods — that is the goods of art and education — is now more significant than ever.

Unfortunately, our recent consciousness of the physical pollution and desecration of the environment that has taken place during the last three centuries, and with alarming swiftness during the last three decades, is still mainly confined to visible environmental results and bodily illnesses and injuries. But we must be equally conscious of the mental pollution and cultural desecration that results from the imposition of our uniform mechanical model on our many-layered historical heritage. Not least we must realise the massive damage done by our own special cultural products — the mass production of printed matter, of pictures, of films, of scientific and scholarly papers no less than the daily outpourings of the mass media. All this has done as much to degrade our minds as our physical conquests have done to degrade the planetary habitat. The excess storage of insignificant information, the excess transmission of unnecessary messages, the passive submission to the constant symbolic bombardment by images and sounds of every sort, culminating in the nerve-shattering extravaganzas of amplified electronic "music" are fast reducing even our genuine cultural achievements to an agglomeration of astronomical dimensions that will be inaccessible to the mind. No system of condensing this bulk or retrieving its separate items will do anything but add quantitatively to the chaos.

Here again there is no technical solution that does not in itself add to the problem. We must call in more ancient forms of technics, the direct cultivation of the organs of the body by new experiments in the well-tried religious practices of mental concentration: fasting, withdrawal, sacrifice, voluntary poverty, to reawaken the full powers of the mind and so further personal creativity and individuation. If we wish eventually to be able to make good use of the computer, we must reserve it strictly for such functions as the human mind cannot by itself perform, or only at a slower pace. And instead we would do well, for example, to cultivate again the art of memory, up to a point that the Greeks and Romans had achieved before writing had become common.

"Do-it-yourself" is the basic injunction for cultivating the mind without obeying the mandates of the power complex, and meekly going through the usual mechanical and bureaucratic channels in order to make use of its elaborate facilities. If we are to have the strength to salvage all that is humanly valuable in the present power system, we must first seek salvation in the individual soul "acting alone or with the help of others," as the Athenians put it in their Ephebic oath. Whereas the first duty of the power system is to obey orders, the first step in shaking free from this system is: Do it yourself! Not merely act for yourself, but see for yourself, feel for yourself, taste for yourself, discriminate for yourself, learn for yourself! Hegel's proud definition of a truly educated man as one who can do anything any other man can do, makes technology itself an integral part of our higher culture.

The first step in the expression of such spiritual independence would be our deliberate rejection of the mythical imperatives of "progress", and our detachment from its sacred abstractions: invention for invention's sake, speed for speed's sake, quantity for quantity's sake, change for change's sake, automation for automation's sake, external control for control's sake, material abundance for consumption's sake. To recover human autonomy we will give a low priority to those processes that the power complex magnifies and specialises in, and a high priority to all those activities that are in some degree within easy personal reach: the local scene, the walking distance, the small community, the intimate face-to-face group, the voluntary services enhanced by families, neighbours, and friends, the modest daily activities of the household, the workshop, the garden, the library, the studio. Only if we turn to the human organism for self-cultivation and spontaneous expression can we enjoy the genuine advantages offered by a resourceful technology without allowing ourselves to be imprisoned within the system itself.

This is not the first time that a mighty collective effort has been needed to overcome the structural defects of civilisation. Such a conscious effort at an inner transformation took place on a large scale in both Asia and Europe from the sixth century BC on, and remained active for a thousand years. This movement was marked by an outburst of subjective creativity, which transferred the seat of authority from the royal and priestly establishments to the small community, the congregation or synagogue or ecclesia, giving respect and leadership to the meek and humble rather than the proud and powerful, embracing even servants and slaves, in a universal fellowship. One must interpret this as an attempt by purely personal effort to withdraw from, if not to dismantle, the megamachine. This emphasis upon personal responsibility and cultural individuation runs through the prophetic religions, the classic philosophies, the more universal ethical systems: from Buddhism, Zoroastrianism, Confucianism, and prophetic Judaism, to Stoicism, Christianity, and Islam.

Nothing like a complete overthrow of the power system was actually achieved or even attempted. Yet this upsurge of conscious religious, ethical, and aesthetic activities left its humanising mark on all the arts, and mitigated some of the worst political and economic evils of civilisation. The attributes of personality, once monopolised by divine kings and their upper class satellites, were transferred, however slowly and grudgingly, to the more lowly members of the community; and in time this democratisation of culture affected technologies, through the shifting of heavy labour from slaves and conscripts to the power machines, the new water-mills and windmills. At length in the organisation of the Benedictine monastery the subjective and objective expressions of poly-technics were reunited as never before in a new type of personality. Work itself became a moral duty and an enlivening spiritual activity; and the regularity imposed by Benedictine ritual upon work laid the foundations, as Werner Sombart pointed out, for the strict mathematical accountancy and mechanical regimentation of the new capitalist regime.

The right questions

There are many lessons still to be drawn from a re-examination of the historical lapses and failures of the axial religions. But their most fatal weakness was something they shared with the power systems they sought to replace. Ideologically, each new religion or philosophy turned out to be a closed system, incapable of absorbing fresh knowledge and experience, and unwilling to modify its subjective postulates, preconceptions, and dogmas, once the institutional forms of the new religion had crystallised. If we are to fare better in rebuilding the moral foundations of today's society, we must embrace every valid part of the human heritage, past, present, and potential. The rebuilding of man's image of himself, which is what the earliest of rituals and the latest of religions have sought to do, is the first step, on this hypothesis, toward overcoming the present automation of automation. No past religion or philosophy will in itself serve this purpose; but an understanding of the ancient practices of religion and philosophy will help restore those parts of the psyche that are now enfeebled, paralysed, or obliterated. Georges Friedmann, in his book *La Puissance et la Sagesse* (Power and Wisdom), has come to essentially the same position I have reached here: a remarkable confirmation from a world authority on modern technics, with a different ideological background from my own.

The underlying premises of this paper, I am aware, run so contrary to those of the "myth of the machine" that my conclusions cannot help being shocking to even those who have been sufficiently patient and open-minded to follow my exposition. What is worse, this interpretation of the interplay of technics with the higher activities of the mind — which in fact leads to a rational appreciation of even the most primitive manifestations of technics — runs the risk of being completely misunderstood. In no sense is this a rejection of the cumulative technological advances from early paleolithic culture to the nuclear age.

So let me clarify what I take to be my essential contribution by reducing it to its barest outlines. In relating technics to every phase of human development, and not limiting it to the improvement of tools, machines, and electronic devices, I have sought to do something that long awaited doing. First, to recognise the objective, concrete, visible, measurable, externally conditioned factors in technics already observable in many other species besides man: witness the works of those masters of technics, the birds, the bees, the ants, the beavers. At the same time I have related such constructive behaviour, such orderly repetitive and rhythmic activities to the neglected but equally significant subjective processes peculiar to man — dreams, apparitions, formative images, emotionally charged ideas, creative upsurges — all of which were dimly in existence, presumably, before any artifacts were shaped by the human hand. If this interpretation holds, some of the greatest technical feats of man were called forth not by a grim struggle for existence, but by a subjective efflorescence in which magic, myth and religious awe played a dynamic part. An adequate conception of technics must do justice to both sides of man's nature.

The views I have expressed here have been presented with ample historical evidence in *The City in History* and *The Myth of the Machine*; but these books have been too recently published to have come under sufficient scholarly scrutiny and appraisal. For all that, at least one anthropologist has clearly perceived the implications of my interpretation of technics and human development; and by the very terms of his rejection he has paradoxically given support to my description of the modern megamachine! I refer to the review of the first volume of *The Myth of the Machine* published in *Science* by Professor Julian Steward. "The thesis of this book", Steward noted, "has inescapable moral and political implications for the contemporary world. If two million years of cultural evolution results from man's mind rather than the imperatives of technology, man is presumably able to devise a better society . . . If on the other hand, economic, social, and political institutions are inevitable responses to mass production and distribution, to what extent can the human mind, or reason, reverse or deflect these trends?"

If this second option were indeed the only open one, technics as practised under the guidance of positive science today would have precisely the cosmic status attributed in the Sumerian King List to the institutions of Divine Kingship. Like Kingship, our technology must have been "handed down from Heaven." If so, not man but the Gods would be responsible for its existence, its enormous power, and its ultimate destination. On such terms the modern megamachine, eviscerated of all latent human attributes except abstract thought, would have absorbed all the attributes of Godhead, and mankind, in Teilhard de Chardin's words, would be reduced to mere "particles", at best specialised cells in a megalocephalic planetary brain. In his framing of this second option, I am indebted to Professor Steward for having restated as a commonplace of contemporary thought the ancient Myth of the Megamachine.

Let me allow the issues I have raised to remain open at this point for your further consideration. If I have not probed deeply enough into the relation of technics to every other manifestation of human culture, I have at least shown that the abstract mechanical model that has seized the imagination of Western man since the sixteenth century is only a specialised substitute — a poor make-shift — for the much richer organic model needed for probing the past achievements and the future potentialities of man. However presumptuous this analysis may seem, please believe that I fully realise how speculative and tentative this paper actually is, and how impossible it is to present even its soundest conclusions, except as a basis for further intellectual exploration and experiment. No one mind, no one generation, no one culture will find the answer to the problems I have raised. But perhaps a fresh start can be made if we dare to ask the right questions.

Computer models for world problems and policy formation

Sam Cole and Craig Sinclair

In Asimov's famous trilogy, *Foundation*, he sketches life in some future century – a society of a million worlds scattered across the Milky Way.[1] The old Empire is crumbling and only through the efforts of Hari Seldon and a band of dedicated psychologists is man saved from many thousand years of despair and anarchy. The most important tool at Seldon's command, barring only the courage of his disciples, is a most mathematical model of the evolution of society during this critical time. Asimov's story is to us not only exciting but credible. It is exciting, not just because it is a good story but to those witnessing what is possibly the embryonic development of Seldon's model in the real world, it is fascinating to speculate on its future growth.

We are referring, of course, to the current worldwide interest in modelling the world, sparked off, in the first instance, by Jay Forrester's book, *World Dynamics*.[2] Whether or not this first attempt at modelling (one does not nowadays need to specify that one is really talking about *computer* modelling) was satisfactory or even sensible does not concern us in the initial excitement. But, even in Seldon's model, set in its plausible distant future there were so many uncertainties. So perhaps we should hesitate at this point and try to unravel our arguments a little and ask a pertinent question. Does Asimov's tale tell us more about the future of computer models than computer models tell us about the future? It is not exactly this question that the present paper sets out to answer, although we will return to the *Foundation* occasionally throughout our discussion, but the more pragmatic one of what we might reasonably expect to get out of our present attempts to build large scale computer models.

The paper is divided into three sections. The first sets out briefly the process of modelling and quickly moves to a discussion of the circumstances in which large-scale computer modelling might be expected to be useful. There are, of course, both advantages and disadvantages to this kind of formal modelling and these depend to a large extent on the purposes to which the results of the models are ultimately to be put. What one expects from an experimental laboratory model designed to illustrate abstract social or economic theory at the present time seems far divorced from a model which is likely to have an impact on public policy making. The second section of the paper goes on to look at various types of 'world model' which have been built or are presently under construction at different institutions around the world. By noting the huge variety of issues which can be classed as 'world problems', the purpose and appropriateness of each methodology is made easier. Some further suggestions as to how the development of large-scale models might proceed and the kind of work we are currently engaged in at the Science Policy Research Unit is described.[3] Finally we return to our earlier question: what are the future possibilities of the world models for policy makers, and we add a cautionary note here.

The potential usefulness of large-scale models

Modelling

Although this might well be the place for a discussion of why we, creatures living in a complex world, should find it necessary to model at all, our competence to do so is severely limited. All we will try to do is to explain simply what we think one is doing during the process of modelling. Side-stepping the question of what we mean precisely when we speak of the 'real world', let us accept that in living out our lives, as individuals and collectively, we receive a huge variety of signals which inform us about our environment in the widest sense. We learn to recognise certain patterns amongst this incredible mass of potential information. These patterns which we perceive and learn to act upon, consciously or otherwise, might loosely be termed 'mental models'. It is these which form the basis of our understanding and form the basis for communication and presumably for our survival. The more signals which appear to fall into the framework of a particular model the more confident we become with regard to that model and our understanding of the processes it describes, and the more reliance we are prepared to place on it in deciding on a particular strategy for our individual and collective action. These 'models' may vary in status from that of 'old wives' tales', art and literature to fully tested, rigorous and internally consistent formal mathematical theories. The former, whilst they are vivid and important empirical observations, can be disregarded for the moment as we look deeper into questions of formal modelling.

Now it is necessarily the case that, although there are advantages in this formal modelling process, there are also disadvantages. Principal amongst these is the danger that if we try to force too much information into a particular mould we may forget many very real and important pieces of information which do not constitute 'noise' in terms of the understanding we wish to gain. Since what we as individuals actually believe we are entitled to 'forget' quite naturally depends on our own personal previous inputs, it is not surprising that there are differences of opinion and differences of culture and ideology. Further, since we must recognise (or at least our current model tells us that we should) that history does not repeat itself in every detail, it is questionable whether a common agreement as to the 'best' mental model can be achieved.

Computer models

Before we move on to examine the use of computers in modelling we must consider briefly the question of measurability.

Given that we recognise some factors in our lives as 'goods' and others as 'bads' and at the same time these exist in neat physical packages, we can simply count their number to have some idea of their relevance to deciding between alternative courses of action which involve only a single type of good or bad. When possible courses of action involve comparing mixes of these goods and bads the choice is more difficult and so we have to invent some common measure to aid us in our choosing. Although we must recognise that the flexible currency system we have evolved is not entirely satisfactory, even for the comparison of the most mundane goods, its impact and usefulness have been tremendous and, in a certain limited sense, it does even succeed in enabling us to compare the most easily quantifiable goods against what is perhaps one of the least quantifiable, human happiness. But this is about as far as it goes, and in fact in most practical everyday choices we are only able to rely to a severely restricted extent on the quantifiable part of our total mental model.

In those areas where measurement has been most easy, in particular the physical sciences, progress with regard to the development of understanding appears to have been aided vastly by our ability to relate together observations made on different scales of measurement, and to define new scales of measurement in terms of the relationships so 'discovered'. But measurement is not the only advantage of physical science. Added to this is the ability to experiment, to isolate and test particular systems of interest, and to create a constructive partnership between the collection of empirical observations and the construction of understanding and theory. However, we should not forget that whilst it has been relatively easy to identify (at least in terms of our current theories) parameters which we believe to be invariant through both time and space, there are many problems which we cannot solve, even though we think we know the underlying theory, purely because we do not yet know how to manipulate the numbers involved.

'Computers' can help to relieve the situation to a certain extent for some problems. If a calculation is straightforward but tedious or repetitive, a computer is invaluable. If a model calculation is 'too difficult' because it involves a large number of variables or non-linear relationships, a computer may be essential, but there are clearly dangers in accepting the results of such a calculation unless they can be understood qualitatively in terms of the principal assumptions of the model.

Between the 'real world' and the numbers which appear on the computer teletype there are several stages, each of which tends in practice to be of different importance to different academic disciplines and each of which contributes in its own way to undermining the credibility of computer models. Of paramount importance to the present discussion is of course the first stage, which we have already touched upon. That is the theoretical model based on our mental model of the real world. Our perception of the 'correct' theoretical model may vary, but before we can proceed further we must nevertheless have one. Getting the right theoretical model is without doubt the most important stage in the development of a computer model. We will comment again on the difficulties of establishing theories in social systems or any other system where either process is non-reproducible.

The next stage in the mathematical modelling process is the construction of the mathematical model itself and this is, of course, the process which is of most concern in this paper. This involves writing down one or more algebraic equations relating the x's and y's which represent the quantities that our mental model tells us are the most important parameters to be considered. At this stage we must decide not only on what parameters are to be considered but how they are to be treated. For example, if we believe 'population' to be an important parameter in our model we must decide at what level of aggregation it is to be treated; if we believe population growth depends in a particular way on certain parameters we must translate that particular way into a formal algebraic equation or its equivalent. At this stage processes which, in the mental model, are assumed to be continuous are also continuous in the mathematical model, but just as the mental model is an abstraction with necessary simplification of the perceived world, the mathematical model is an abstraction and simplification of the mental model. Following sections will discuss various developments in large-scale modelling in terms of their theoretical and mathematical properties but for the moment we move on to consider the last and often forgotten sources of error in the modelling process.

The reasons they are forgotten are twofold. In the first place those disciplines concerned with trying to understand the 'real' world tend to concern themselves with those aspects of the modelling process we have just described. In the second, if a computer is used to analyse a theoretical model, the programme, be it in the form of a ready-made 'package' such as a multiple linear regression or any other, is treated as a 'black box'. The workings of the 'black box' do not concern the user, and he tends to assume that it operates with absolutely no error. This simply is not the case and it is perhaps sufficient to say that there exists a whole discipline, 'numerical analysis', which is largely concerned with removing the uncertainties and instabilities which can arise in the mathematical and numerical representations stages of a digital computer calculation.

Feasibility and utility

The Science Policy Research Unit has engaged in a feasibility study for a European socio-economic model. The ultimate purpose of the model would be to examine long-term European policy issues in the light of world and regional issues, eg, those of industrialisation, energy and other resource supply. Clearly the same considerations of feasibility are likely to apply to any large-scale economic model and so we repeat them here. The principal component of any feasibility study is examination of the eventual utility of any successfully completed activity. For the contemplated modelling to be helpful to policy making, it must above all be clear and shown to be based not only on well-understood and acceptable representations of the mechanisms at work but these mechanisms must represent sufficiently a summary of the main influences and consequences of the system studied. Put another way, the results must be communicable, at a sufficient level of resolution to be relevant to practical issues, and to be acceptable in that they do not lean upon controversial or over-speculative views about the mechanisms at play. It should be emphasised that their role is not to supplant all other inputs into the decision process, but rather to supplement the various perspectives brought to bear on practical policy issues.

However, a proposed model which would be useful if it were available may not be technically or logistically feasible. Technical feasibility requires both the availability of the computers needed to cope with the complex and detailed models of the type being considered here and, of course, a methodology capable of handling representations of dynamic and interactive systems. Logistic feasibility requires both that the underlying theories upon which the model is based are individually sound and mutually consistent and that coherent data compatible with these theories are available for the European zone. The logistic feasibility must take into account the present availability of relevant data, gathered as they have been for the most part for other purposes, and the long time delays and large resources which would be associated with the gathering of new data. Further, since both interdisciplinary and international cooperation are prerequisites for the work, an examination of these must be made.

The potential utility to be gained from the building of large-scale socio-economic models such as the Europe model and the variety of uses to which these models may contribute will now be described. It is only possible to evaluate the technical and logistic feasibility of models against an appreciation of some scale of their potential usefulness.

At the lowest level, all formal models help a researcher to clarify ideas through forcing a resolution of inconsistencies or non-coherence in the previous set of theories, which evolved often for different purposes and was based upon data gathered independently. At the same time, the posing of the needs of formal models for unambiguous representations of exhaustive but exclusive sets of theories facilitates more precise communication both between researchers in the same group but of different disciplines, and between different groups of researchers in their own institutions. The first two levels have considered the utility of modelling for the genesis of a corporate body of theory. Related to these are questions of the data required to substantiate that body of theory. The wider the agreement on the body of theory to be substantiated the more coherent and consistent is the data-gathering activity. Formal modelling is a specific catalyst for this. Policy makers derive benefit from having a generally well-substantiated background of theory but this does not necessarily imply that a formal model must exist, since there is a variety of channels through which the theoretical background for policy making can be improved. However, where there is a direct channel for communication, its existence can bring into play a better understanding and perspective of the environment in which policies have to be made. At the minimum, an accepted model, though necessarily an over-simplification, can act as a useful cross-check for considering the ramifications of a proposed policy. Again this does not presuppose that we can reach the highest level of utility at which there exists a model which is a precise tool for evaluating that policy.

Some current developments in world models

Various approaches to global modelling are currently under way at research institutes through the world. The present discussion will restrict itself to a limited number of computer models although this is not intended to denigrate other models or methodologies. The Science Policy Research Unit is currently being funded by the UK Joint Research Councils (Science Research Council and the

Social Science Research Council) to examine thoroughly all these programmes. The project is entitled 'Application of Dynamic Modelling and Forecasting to World Problems' and is intended to provide a basis for decision as to future work of this nature to be sponsored in the UK. The International Institute for the Management of Technology in Milan, Italy, represents one of the first institutional responses to these problems as they affect the management of technology which will utilise the essentially systems analysis approach inherent in modelling techniques.

The discussion will use Forrester's original model World 2 as a focus and consequently a short description of his work will be made. The parameters explored by Forrester's model are, with a few additions, the same as those of the 19th century philosopher-economists, Malthus and Ricardo. Forrester's conclusions are generally 'Malthusian' since the message of his work is that we live in a finite world and that with rising populations and living standards an end to growth is not only inevitable but imminent if present trends continue. He concludes that growth will end through the combined effects of rising populations and wealth coupled with declining natural resources and geographical space and the rising hazards of industrial pollution. Various 'modes' of collapse are discussed − all characterised by a rapid decline in population levels. Forrester is careful to point out that his model is not sufficiently precise to predict the actual mode of collapse which would follow from a continuation of current trends, but argues that there is no simple solution which avoids one hazard without encouraging another. He suggests that catastrophe can only be avoided by implementing combined and drastic policies with respect to the control of population and industrial growth and that we should seek a state of world equilibrium characterised by stationary levels of global populations and wealth. He actually argues that social collapse would probably anticipate a physical collapse of industrial society, so the parallel with Asimov's tale is almost complete.

Whether or not confidence can be placed in the actual results of Forrester's model (the conclusion of the Science Policy Research Unit's analysis of the work[3] was that it probably shouldn't be) it provided a stimulus to a large amount of other work on large-scale modelling which aimed to examine a variety of 'world problems'. Under this title might be included both genuine worldwide problems and local, but widespread problems. The latter, which possibly form the majority of world problems, are those with common causes and symptoms but they can in principle be resolved independently in each area. Amongst these one might include over-population, whilst in the former should be included certain types of pollution of the oceans and atmosphere. Secondly one should distinguish between the time-span over which an issue is acute and whether a particular short-term local problem has a tendency to become a long-term widespread problem, an assumption which underlies Forrester's world model. The models considered here all have a strong social element (ie they are not conventional, ecological or economic models) and embody concepts drawn from many disciplines.

Interactive models

Models can be described on the various spectra which display their properties. The first of these is perhaps their degree of interaction with 'mental models', that is whether they are interactive or merely trend projections. In principle any model is interactive but the ease with which this takes place is variable and has been explicitly built into some modelling methodologies. Mesarovic and Pestel have built one such model.[4]

> In such an operation the analysis is the outcome of both logical and computing capabilities of the machine on the one hand; the intuition, experience and heuristic capability of man on the other. Such a symbiosis avoids the pitfalls of relying solely on the computer for policy analysis, which by necessity leads to a mechanistic view of the situation; it provides also means for a creative use of computer techniques for extending the logical capability of man in long-term planning and analysis while preserving the ultimate responsibility for prediction, planning and decision making in his hands.
>
> Essential in the interactive computer simulation (analysis) is the following: the computer contains a representation of dynamic processes relevant for the situation (problem) of concern, the alternatives available to the decision maker for affecting the situation, and the constraints which he must observe in implementation. The decision maker observes changes during the evolution of the system in time (via an appropriate set of indicators) and by using his own judgment determines when a corrective action is called for. Depending upon the specific aspects of the undesirable behaviour, the decision maker asks the computer for more detailed information on the situation, the type of policy alternatives, strategies and measures available, the constraints to be observed, etc. He then makes appropriate decisions and instructs the computer accordingly. As a rule, the interaction between the decision maker and computer is more complicated, involving numerous iterations before a final set of decisions is made. However, the basic division of labour between man and machine is always the same: man decides on values, priorities, costs and the level of risks to be taken; the computer indicates the breadth of choice and likely consequences.

In a sense this is an ideal and the goal which tends to be sought by computer modellers. Whether or not this goal is too ambitious we have to recognise that, as pointed out earlier, theory is in general extremely weak for many of the non-physical relationships currently included in global models, and data are poor, even for many physical phenomena. In any case, it is pertinent to question whether effort should for the moment not be concentrated on the development of satisfactory theories in an attempt to explain past phenomena. Although the Mesarovic–Pestel model possibly goes furthest towards recognising the complementary advantages of man and machine it is perhaps least suitable for the development of theory. One could argue that a straightforward 'projection of current trends' model such as Forrester's, or Meadows' World 3 elaboration of this model, is the best approach to theory testing against historical trends.[5] Once the ability of such a model to mimic historical 'reality' has been established it can easily be converted into the Mesarovic–Pestel model since virtually every relationship in a Meadows type model is flexible to 'policy' given a sufficiently long time scale. So degree of dependence on established theory is the second model spectrum.

Goal-oriented models

Intermediate between trend-projection and interactive models can be seen 'goal-oriented' models and 'scenario-oriented' models. Examples of goal-oriented models are the South American Bariloche models and the work of the Delft group in Holland. The Bariloche team is examining indexes of welfare in an attempt to determine the minimum amount of physical goods which are necessary for human beings to be happy. They start by asking what is necessary for a tolerable life but recognise that the physical necessities will not be the same for all people. The essential elements for human life are then considered in the light of various technological alternatives for achieving these elements. They intend to include their index of welfare into a model of humanity compatible with the ecosystem. This would be done by setting an egalitarian objective for the model to be achieved in the shortest possible time consistent with social and physical constraints. The model takes into account another factor which will be considered again later. It recognises differences between levels of development and wealth across the peoples of the world.[6]

A slightly different approach to goal-oriented modelling is used by the Delft group.[7] Rather than setting an objective to be achieved at the earliest date, their model seeks to maximise a total welfare function over the next hundred years or so. Whilst it cannot, of course, determine the political *values* which decide what heritage of problems current society should pass on to its descendants, it can be used to demonstrate what are the possible consequences of particular policies with regard to resource utilisation and the like. Again this presupposes that theory and data are at a satisfactory level. The Delft team have initially used a modified version of Forrester's model for their experiments and constructed composite welfare indicators from the material standard of living, pollution and other factors embodied in that model.

An alternative approach selected by another Dutch group led by H. Linnemann is to take a given world scenario, in this case the 'Doubling of Population' and to examine the possibilities for achieving satisfactory living standards in the context of that scenario. The study uses a 'policy model' incorporating a number of 'instrument variables' that will allow the realisation of certain targets of socio-economic policy. These targets are at first sight not dissimilar to those of other groups:

> The aims of the world community will comprise (a) a decent material standard of living for all people; (b) a reasonable reduction of disparities in income levels between nations; (c) maintaining or improving the life-supporting capacity of the world (the biosphere); (d) the responsible use and management of non-renewable resources.

Whilst it would be unfair to single out any team for criticism, since most groups, including Sussex, have similarly general and vague 'objectives', it is clear that there is a need to state more precisely what one means by, for instance, 'reasonable reduction of disparity' and exactly what risks one is prepared to take with regard to the long-term future of mankind in order to alleviate the troubles of today. Again these clearly depend on political values — and where models might be able to help is in bringing into perspective the results of selecting a particular set of values. For the moment one need hardly say that there are probably widely different views between the objectives of the developed and underdeveloped nations in this respect. What the developed countries fear for the future the underdeveloped countries might claim for themselves today. On the other hand the objectives of formal institutional approaches such as that of the International Institute for the Management of Technology to the narrow sectorial areas are defined in terms of official views of the methods of problem solution. As the need for careful overview remains such, they offer partial responses and sometimes piecemeal answers.

Clearly a major and valid criticism of the Forrester model was its extreme level of aggregation of all variables in that it did not distinguish between the various world sectors from a social, economic or physical point of view. Not only does this hide serious issues arising from heterogeneous distribution of both problems and resources throughout the world, but causes the models to be unsatisfactory aids to the development of policies. For example, if it is true that, as Forrester suggests, supplies of non-renewable resources may be exhausted within a relatively short period due to their intensive exploitation, then restrictive world policies for consumption are needed. But if we first recognise that 80-85% of the total non-renewable resource consumption is by 25% of the world's population we can recognise immediately where the restrictions would be most effective. We further recognise that a model which does not take account of world heterogeneity may be worse than useless and policy recommendations based on it may have no possibility of being accepted by the majority of mankind. Clearly this has been recognised by many of the groups so far described and each has attempted to resolve the associated theoretical, empirical and computational difficulties. The Japanese group led by Y. Kaya has made several imaginative attempts at this problem. Members have examined in particular the threshold for economic aid from the developed to the underdeveloped world to permit the latter to 'take off', and the relationship between resource depletion and world economic development. They make no claims for the results but it is interesting to note that their preliminary results indicate that a threshold does exist and that present levels of aid fall well below it.

A project with a similar aim, although restricted to the intra-national situation, is that of the International Labour Organisation on the *Bachue* model (so called after the Columbian goddess of love and fertility and of harmony between nature and man). This model is designed to test the interactions between population growth and employment in order to throw light on employment policies in developing countries. Although this model is not a world model, the problems it tackles are important to the majority of the world's people. The model may help to answer questions such as whether a developing nation should invest its capital in advanced technologies like nuclear power, or in agriculture or 'intermediate' or 'appropriate' technologies. Another possibility is that the model will clarify issues related to birth control or rural-urban migration. With regard to the latter, the current 'closed city' debate and its effect on long-term national growth is examined. One distinctive feature of this model is its high level of disaggregation (another modelling spectrum); for example employment is divided into 15 sectors including modern and traditional agriculture, modern and traditional manufacturing industry and employment in service functions. To what extent a model should be divided is still a matter of debate amongst modellers. Of course, in theory the problem under consideration ought to determine the 'upper limit' that should be used in a first attempt of a problem. However, it is possible to exaggerate local problems into world problems, and in absence of satisfactory theory such a model can be misleading. On the other hand a finely sub-divided and comprehensive model, although it should, at first sight, have the detail necessary for policy making, is almost bound to suffer in part from lack of substantive theory and data.[8]

Quality of life

Finally we return to the question of welfare indicators and their relationship to models as a whole. It goes without saying that for policies to be viable they must not only be physically *possible*, they must also take account of changing social environment. Consequently straightforward projection or optimising models may be unsatisfactory tools since in reality social inertia (static or dynamic) can present obstacles to necessary policy changes as large as any physical resource shortage. Forrester's model contained a simple and much criticised 'quality of life' indicator. Several groups are examining this factor in greater detail among which the Japanese group has a programme of work under way to look at the 'value standard of the Japanese people and its change'. The Sussex team is similarly involved in both a national and a worldwide study. One Sussex approach to gaining an understanding of the way the values of today affect the future is based on the assumption that there are fairly strong deterministic relationships between demographic, economic and environmental conditions and people's traditions and aspirations. It is interesting to note that Seldon's model too was based on the collective psychological forces of vast populations throughout the universe.

The view is then taken that in the long term policies will largely be determined by the social pressures populations exert on policy makers as they attempt to increase their individual and collective utility. These social pressures are continually modified as the relative urgency of different needs adjusts to changes in the physical and social environment. The explicit representation of value changes within a model can reduce the chance of evolving socially impossible policy programmes from the forecast trends. At the same time utility parameters in terms of an 'aspiration gap' (the difference between what people want and what they have) provides an important social element for a welfare objective function and a possible 'driving force' behind the development of social processes within a dynamic model. To what extent such a system may be formally modelled is debatable but as we have noted earlier a framework such as this provides a valuable tool for structuring the research effort.

We have carried out a literature survey focussing mainly upon quantitative research and especially on those studies considered to be more amenable to experimental modelling. A number of conclusions can be drawn from this review. The processes of social change, both at the level of human experience and at that of political policy, are comprehensible within the terms of our framework and established theories of social phenomena. However, while the actual features of these processes emerge, quantifying the causal relationships involved is difficult in view of the lack of appropriate research in many of the areas concerned. Any attempt at building a dynamic simulation model at the present stage would be forced to rely on a great deal of approximation, of inferences from cross-sectional relationships to time series effects and so on.

A European model – some practical issues

As explained earlier, the Sussex group has carried out a feasibility study for the construction of a European model at a level of description capable of analysing European policy issues as influenced by regional phenomena. As pointed out, certain logistic and technical problems beset the potential modeller and some of our conclusions with regard to the European model are equally applicable to other large-scale economic models.[9]

For the construction of large-scale models, attention has already been drawn to the need for a more coherent body of theory, of data, of cooperation between research and policy-making institutions particularly in regard to the coordination of activities. Foremost among these problems, however, is placed the dearth of a well-accepted corpus of theory. Even if one rejects the view that there could be a single unique and comprehensive theory of sufficient simplicity to encompass the wide variety of phenomena combined in comparative socio-economic studies, the problem is not resolved. For example, it is pertinent to consider the situation with regard to theories of social change. In the broadest sense, theories about social processes, though varied, may be reduced to a small number of variants although, in practice, few people rely on one variant.

Theories about 'industrial society' draw on four variants. Briefly, these are conflict theories (the most important of which is Marxism), functional theories, ie those derived mainly from anthropological sources, interest theories which revolve around market mechanisms as the principal forces in society, and finally hierarchical theories in which economic growth is a major mitigating force in reducing inequality in an accepted social and economic hierarchy. Since every theory has been formulated to deal with a particular set of problems, often in a particular historical setting, one should question at the outset whether it is sensible to construct a model which is holistic not only in the social and economic sense but also with regard to ecological and environmental factors. The 'holistic' argument is that it is better to include all effects which are believed to operate within a total system, however poor the current knowledge base with reference to each phenomenon. Nevertheless, each theory has a role to play and the model and its environment should be so organised as to make possible the maximum use that the varying perspectives afford by broad eclectic philosophies. If theory was in a uniform state of advancement, then the holistic approach might prove to be the potentially most rewarding in the short term, in its capability of drawing attention to the complex interrelations of problems. However, the high level of complexity which would accompany a high resolution holistic model could well have the effect of clouding and possibly confusing other and possibly more urgent issues. Where the current state of theory is also uneven in its 'load-bearing capacity' this can result in the uncertainties which arise from the less well-evolved theories distorting or denigrating the precision of the better established theories.

For the purpose of the European models, it can be argued that a balanced solution to this issue offers the better hope for advance. The approach suggested is to construct an experimental model which takes into account the most important social and economic determinants of long-term development and, in particular, concentrates on those issues for which there is the soundest theoretical base. Such a model would, itself, be used to 'drive' a set of more experimental sub-models representing other important but more speculative and therefore, so far, less well-researched problem areas. The advantage of this approach is that it permits best use to be made of available theory and at the same time avoids the hazard of introducing too many tentative relationships into the main body of the model. One question bearing discussion at this point and one which has been referred to several times previously, is the vexing question of the appropriate level of detail within the model. One has to strike a balance between the level appropriate to the needs of policy makers, the capabilities of computers and the availability of data and the state of the theory. Macroeconomics, microeconomics, and regional theory all bear at different levels on slightly different aspects of reality. What one strives for are levels of description capable of accommodating regional differences without unnecessary proliferation of detail.

In considering the data base that is required to calibrate such models, a difficult but not insurmountable problem is faced. It is true that vast quantities of data and information concerning many demographic and economic variations and developments have in the past been collected in a systematic fashion as far as individual statistical series are concerned. It is also true to say that even at the national level these data are of varying quality and cross-comparability is hampered. For models of the type considered here, which essentially concern interaction, the individual series ought to be coherent in relation to each other.

The problem of calibrating a model using current data by considering the following types of statistics can be exemplified as follows: at the international level, comparative data are available at a highly aggregated level, are chiefly economic or demographic, but are inevitably delayed and often slanted by the particular requirements of the organisation supervising their harmonisation. At the national level, data are more up to date but the methods of supervising organisation vary from country to country and comparison is more difficult; at the regional level data suitable for such representation as input-output matrices are becoming available, but flow data between regions are of varying quality. With regard to social data even in the simplified form of social indicators, comparative data of other than straightforward demographic detail are sparse. National surveys of social phenomena are often carried out on a regional basis, but their value as data for regional characterisation is usually limited. It has been our experience that this reflects a mosaic pattern where each piece has been gathered for different purposes and in a form which reflects a need peculiar to the theoretical underlay of a national culture and the statistical purposes of the data-gathering organisation. Our contacts to date with institutes located in other countries suggest that this will be the case generally. For example, country comparative statistics of depreciation of capital goods are heavily contaminated by the varying fiscal arrangements of each country. Nevertheless, by facing this issue as posed by formal modelling we would anticipate a further stimulus to the current work of economists and others in this field. At this point attention is drawn to the fact that long-term models require a data calibration approach which reflects the long-term trends in those data, uncontaminated by short-term variations.

The two other aspects of large-scale modelling are, first, the questions of cooperation within and between research organisations, and between research organisations and the sponsor, and secondly the overall coordination of the modelling programme. A continued contact between the bodies is a prerequisite for a successful programme. What is also required is effective coordination. Agreement about such specific issues as programming languages, nomenclature and choice of variables are but one type of issue for which agreement is essential. There is also a need to establish a work rapport between a general philosophy, the underlying theory, the structure of the data base, and the goals of the programme. One conclusion of our work to date is that the best of intentions in this respect must be accompanied by firm commitments to exchange of information and possibly short-term exchange of personnel.

Finally we will re-examine the potential utility to be gained from the striving for large-scale socio-economic models. It has been our experience on the European model that the necessary construction of a set of consistent mechanisms to describe a range of associated phenomena has led to a useful clarification of ideas within the group. Besides this, the posing of formal models to researchers both of different disciplines and from other research groups has undoubtedly facilitated more precise communication, and we question whether such a level of communication between researchers of very different backgrounds could have been achieved as readily in any other way. An appropriate approach of modellers to policy makers is discussed at length below in the section on models and policy.

The search for data presupposes that many of the mechanisms which the data are intended to calibrate are understood. It is an unpalatable fact that this is seldom the case. Given the present uneven level of understanding of many of the processes which are obviously involved in the working out of socio-economic processes, it is probable that large-scale models, although not yet of high utility directly for policy purposes, would nevertheless be of considerable value to further understanding of the general forces and their interactions affecting specific issues. In particular it would demonstrate those areas in which policy decisions cannot be confidently made without further research into either the collection of data to support existing theories or the evolution of better theory.

It might therefore be argued that the aim of current modelling work should be the development of a modelling strategy which permits best use to be made of already established theory and which takes maximum advantage of modern dynamic modelling methodologies within the context of the policy-making process. In other words, one is seeking to maximise utility by seeking the right combination of technical opportunities whilst recognising logistic constraints.

World models and policy making

The relationship of modelling to policy making is not yet clearly defined. A quotation might be a suitable entry into the following discussion of the relationship as it will set, perhaps, the tone that is to be used in discussion.

> The only way to learn the rules of this Game of games is to take the usual prescribed course, which requires many years; and none of the initiates could ever possibly have any interest in making these rules easier to learn.
>
> These rules, the sign language and grammar of the Game, constitute a kind of highly developed secret language drawing upon several sciences and arts, but especially mathematics and music (and/or musicology) and capable of expressing and establishing interrelationships between the content and values of our culture; it plays with them as, say,

in the great age of the arts a painter might have played with the colours on his palette. All the insights, noble thoughts, and works of art that the human race has produced in its creative eras, all that subsequent periods of scholarly study have reduced to concepts and converted into intellectual property – on all this immense body of intellectual values the Glass Bead Game player plays like the organist on an organ.[10]

Are world models from the policy maker's viewpoint, then, no more than a kind of intellectual's game or are they capable of being used to provide policy makers with help in policy making?

The tone of the following, it will be apparent from the choice of quotation, is cautiously sceptical but not cynical. Another name, another game, at this stage of the development of the scientist as analyst of the world's problems, will do more harm than good if presented as the definitive method of analysis. Modelling and policy making will come together by mutual attraction if the claims for the former are not pressed too stridently.

Wittgenstein's entry into the ultimate aridities of his philosophy was the realisation, upon seeing model cars being used to demonstrate a real happening, that pictures and models constituted a reflection of reality which was manipulable. He further realised the enormous potentiality of models and analogues in misleading people by allowing the imagery to defeat the rational calculation. The image he conjured up was of a string tied round the Earth's Equator. By the addition of an extra yard to it the string may, in the mind's eye, be lifted off the surface of the Earth.

Most people, he found, when asked how high from the surface could the string be raised in this way answered an infinitesimal amount! A simple mathematical calculation will give a different answer. It won't be quite enough for a Limbo dancer to get under, but nearly. Models – images – mental reconstructions of reality – are to be treated with care. However, responds the modeller, the computer takes care of that.

Are modern world models capable of giving us useful reductions of complex realities? Before tackling the problem in terms of descriptions of the possibilities of using computer models for political decision it should be recognised that in approaching the subject we are necessarily dealing with a clash of cultures, or rather cultural attitudes. The above example is of the fragility of even simple mental pictures as a guide to problem solution compared with mathematical calculation. It is difficult to discern in it any in-built social, political or psychological bias except that of the intuitive against the mathematical. However, when we deal, as we do in *World Dynamics*, with rather wider aspects of world description it is clear that biases of all kinds may enter. William Kapp, an old campaigner against environmental disruption, says 'this disruption is the outcome of a complex process of interaction of social and physical factors which cannot be adequately analysed in terms of the concepts, theories and perspectives of any of the conventional disciplines'. World modellers are thus a new breed of men whose expertise lies in new areas of thought and as such they clearly carry an enormous burden in proving and defending their usefulness against entrenched opinion. So much the more then must they ensure that their activities and results are convincingly valid, not only to their peers but to the world at large.

The scientific and technical community is only thinly represented in the parliaments and the decision-making levels of the civil services and companies of the world, at least as measured by the number of members with training in this field. Only recently have education and culture come to include mathematical expression, through science and technology, as well as literature and the humanities. Company long-range planning, let alone industrial sectorial plans and national plans, is also of comparatively recent origin, and by no means universally accepted in principle or in practice. The implementation of such plans, indeed the entire concept of planning, is scientific, mathematical in origin and taken to its logical conclusions becomes all-embracing and holistic. The traditional social and cultural attitudes to management, be it of the national economy, industry or the individual company, are based upon the historical educational values and language patterns and assumptions of a society of a radically different technological base. In the UK, 'Locomotives, ships and bridges were the real Victorian gods', says Alex Comfort in *Nature and Human Nature* and he goes on to claim that 'their success is an eloquent illustration of the way in which emotional needs succeed in squeezing into wholly practical contexts when we fail to express them deliberately'. The current technological achievements of our technical industrial revolution would appear to be, at the summit, Sputnik and the landing of a man on the moon, the nuclear reactor and the nuclear bomb, the conscious and unconscious goal associations and functions of which seem often to be more irrationally than rationally derived. Bridges and locomotives, if not the capitalist drive for wealth and production which generated them, resulted, given some drawbacks, in achievements which are generally worthwhile for their own sakes. That is, they had an immediate, rational use and fulfilled an emotional need. The modern harnessing of technology may be much further removed from man's rational use of his knowledge, and nature's resources seem today to be accompanied by the possibility, at least, of larger-scale catastrophes than previously. Further, many of the 'achievements' may themselves by questionable. Thus, it may be that neither rational nor irrational, utilitarian nor emotional functions are being fully satisfied by the modern technological drive. Such a mismatch between hopes and aspirations and results and reality may mean that we are truly at a cross-over point in cultural development and are thus searching, as a society, for new languages, values and methods of expressing these.

The question is, are the mathematical models of the scientific culture refined enough to oust or even supplement the older and perhaps more intuitive methods?

Predictive models

For all that it may be discouraged by their protagonists the element of modelling which will be and is seized upon by those faced with a problem or set of problems requiring decision is the predictive. Indeed their attempts to disavow this aspect of their work has a hollow ring. If they are not making predictions what are they doing? Nagel in *The Structure of Science* defines science, acceptably, as revealing 'repeatable patterns of dependence' and it is this 'repeatability' based upon discovered relations or dependences between man and the material world which allows the claims, for example, of Economics as a science. And it is principally the economists, with the largest history of attempting to build predictive models in sectors of this area of man-material interaction who are most sceptical of attempts

79

to widen the scope to the world as a whole. It will be instructive then to examine what might be demanded by an economist of a predictive model.

Predictive models should, following Heilbroner, be such as (a) to allow the formulation of higher-level hypotheses and (b) to establish useful categories of generalisation. To take this latter first, a 'taxonomic' element is a necessary outcome of the science of modelling since taxonomies in the natural sciences have in terms of structures proved useful in providing visible analogies and have raised fruitful inquiries about function and evolutionary ecology. In the social sciences where world modellers claim to operate, however, the structural taxonomy approach may be less useful since the statistics of physical structure, capital, labour, pollution etc tell us little about the origin and development of the social structure described in these terms. World modellers appear to have no more success in finding the lasting and important elements of social structure and hence social change than have other less ambitious analysts. Indeed, current attempts at world modelling have been almost startlingly bare of any attempt to incorporate sufficiently even the modicum of material that is known about human behaviour in particular situations of interaction.

On the other hand the formulation of higher level hypotheses about, say, human response to the threats evidenced by the results of current models would have been a strong point in support of the models. However, the reverse has probably been true. Such generalisations on this theme as have been drawn have been more reiterations of the necessarily naive and simplistic theories which were subsumed in the original data and relationships chosen. The essentially tautological nature of mathematics — which consists in deriving from a set of axioms or postulates all the consequent relations — has been even more transparently demonstrated by the mathematical models of the functioning of some of the more readily perceived operations and activities of the real world. If it is argued, as argued it appears to be, that in designing a model a socio-political stance must be taken, then the predictions made must be viewed by decision makers with the knowledge of the nature of the stance. A Malthusian base will give Malthusian answers and only the timescale can be claimed to be particularly pertinent. Thus the apologists for such models claim time as the output of final special relevance. That is, given the world's resources, in whatever category, are finite, and no-one argues with this, then the output of first importance of a model designed to calculate the rates of usage must be the time at which decisions must be taken to arrest the process. The embracing of the mystical religions of the East —

> The opposite concept of man is also an ancient one, but it is more closely related to the Eastern religions than to the Western ones. It assumes that man is one species with all other species embedded in the intricate web of natural processes that sustains and constrains all forms of life.[11]

by the most famous of world modellers is not without psychological significance in this respect. The concept of time in these religions is hardly one which fits easily into the strict, logical scheme of machine computation.

It would be more instructive for the world modellers to expand, say, upon the Buddhist theory of reincarnation in relation to cataclysmic events than simply to mention 'intricate webs', if we are to accept this particular philosophical stance.

In order that deductive logic might be used for prediction, given the foregoing, it is clear that some overriding, or overarching hypothesis must be used. It should be used to blend together the fine details of time spans, behavioural uncertainty, and knowledge of the nature of technological change, which must be the inputs to any world model. Both Marx, and earlier Smith, made heroic attempts to describe the nature of technological change which might be claimed each in its own way to have been unsuccessful. It is this last consideration which brings us to an important desideratum for world models, the need for a historical perspective. The historicism implicit in the current models may be, in small measure, comforting, in the sense of providing continuity. But the constraints of present action arise not from an inevitable flow of historical trends but from the views that men hold in their heads of what history implies and these cannot be described in material categories.

Decision making

It may be useful to look at some descriptions used by political analysts of a traditional type of the decision-making process and attempt to see if the output of present world models can be fitted to them.

A simple and straightforward view characterises the political decision-making process as consisting of (a) a search for goals followed by (b) objectives, formulation, (c) the selection of alternative strategies to achieve (b), and finally (d) an evaluation of outcomes. Current world models operate in just the reverse manner. The world system is conceived in terms of existing policies and trends as the modeller sees them. The programme then proceeds in a mathematically logical way to arrive at seemingly inescapable conclusions. The outcomes which are derived are then ascribed to the ineluctable operation of the existing policies and this static description is then bolstered and defended by democratic appeal for better decisions about national or international goals. Since the fine details of the processes are hidden in the mathematics and moreover in the data chosen for the model, the modellers can appear as rational men overwhelmed by the ignorance of the political decision makers who have decided the goals. Thus one major objection to the current operation of world modellers is their belief that goals indeed are or can be given as simple statements. They should be, they claim, the goals of a rationally articulate majority. However, goals inasmuch as they can be simply classified and separated, are the outcome of conflicts of interest, bargaining, and the trading off of one set of wants or needs with another. The political system is the process for arriving at the imperfect set of objectives. World models should, but don't, take into account the 'time span and behavioural uncertainties' of this process. The foregoing discussion will have indicated the extent to which the uncertainties may be incorporated into a useful model.

It is just this implicit, and dangerous, belief of the world modellers about decision making as it refers to the selection of criteria which concerns us. If the welfare of more than one individual is at stake — the undernourished peasant as well as the Washington commuter — then we are in an area of perennial difficulty, that of the intercomparison of personal utility. It would be churlish to suggest that world modellers have to get round this difficulty by the device of making the disbenefits of current policies and trends so enormous by their choice of data that there can be no interpersonal differences in response, but there is an element of this in the presentation of the results. The catastrophe facing us is such, they claim, that these differences are negligible: let us unite against our common fate.

In general terms this is the central problem of democratic government. At the particular level of specific cases it is the cause of much of the dissatisfaction with the performance of government today. Classically, if one may use the word of a situation only a few years old, the problem is that, for example, posed by the siting of the third London airport. It was a problem under review for many years and the subject of official examination, public enquiry and parliamentary debate. In spite of this it has not been possible to reach agreement even after the work of the (Roskill) commission and its rejection by the public. Questions of technical estimation apart, numbers, noise levels, cost etc, the crux of decision making is the varying valuations of the affected people and the problem of reconciling the losses of the disadvantaged against the gains of those benefiting. Does modelling avoid these problems?

This central problem of policy making has been tackled by welfare economists by methods which have finally evolved into cost-benefit analysis, which Roland McKean has characterised politely as 'a method for quantifying the unquantifiable or a sure road to Utopia' depending on one's viewpoint. Such analysis does not solve the problem; it simply renders the decisions more explicit and puts apart those items often ignored in decisions.

Effective decision making is also obstructed by a general problem; it seems to us left unattacked by many computer models, that of human behaviour. Group or personal decisions involved, with directly expressed preferences, can give answers which are not unequivocal with respect to ranking of alternatives. Condorcet recognised and expanded the paradox of a listing which prefers A to B, B to C, and C to A, many years ago. What does modelling make of these problems with its inevitable reliance on aggregation?

If wider consultation is proposed as a solution, rather than say mystical contemplation, then it must be recognised just as clearly that taking group decisions will involve a trade-off between the benefits of this wider participation and the costs of securing them. The very urgency with which the world modellers press their case shows an underlying sense of the trade-off involved in getting their apocalyptic message to sufficient people and the likelihood that the response will be effective.

An expansion of the above-listed stages of decision making as:

- the determination of objectives;
- the definition of the problems that need to be solved to achieve these objectives;
- the search for various solutions that might be offered to these problems;
- the determination of the best or most acceptable solutions;
- the securing of agreement that such solutions should be implemented;
- the preparation and issue of instructions for carrying out the agreed solutions;
- the execution of the solutions

shows that every step is involved with a political or administrative question. The models on the other hand are exclusively concerned with material relationships and they subsume the important social relationships underlying, for example, the exchange of goods and resources.

What are required by policy makers are improved methods for completing each step of the above progression rather than mechanistic demonstrations that the present systems are no good. The relative failure of the systems approach as applied to government decision making does not mean that such methods are always to be doomed to failure. Rather they should act as a stimulus to better and more cautious attempts.

To join in the game for a while, without, to continue the earlier analogy, saying how long a piece of string is, Figure 1 shows the interrelationship and linking of the steps together. When even this simple scheme is organised for satisfactory computer solution we shall be some way along the road to effective decision making about the wider issues.

The views that may be taken of the nature of the problem, as laid out above, can be classified in the following way:

(a) a normative attitude expressed as a statement that it is the cultural values, ie the mentality of people which will ultimately lead to a solution or not;

(b) the causal, Marxist approach which concentrates on an identifiable cause for the crisis. In the simplest view the problem is the capitalist exploitation alike of nature and labour;

(c) the technocratic solution is solution by specialists; each group putting particular emphasis on a particular facet, ecological, medical or economic;

(d) a subspecies of (c) is the managerial approach, the crisis being seen simply as a decision-making problem.

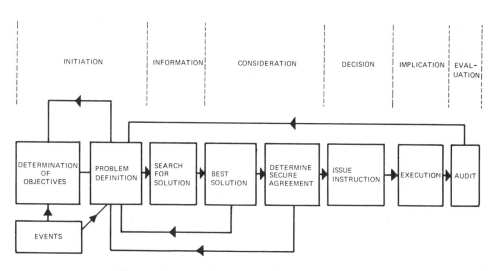

Figure 1 Schematic interrelationship in decision-making process.

It will not take a genius to guess that the approach advocated here must claim to be made up of an integration of these approaches. Note, however, that solutions or even the path to solutions are not being clearly defined. What is being proffered in this section are a few tentative but considered hints as to the kind of stance that must be adopted in treating apocalyptic views of the world's problems. Returning to Wittgenstein, it is the nature and kind of question posed that are important and are often more important than the answers.

Knowledge flowing from experts enclosed perhaps in the paradigms of categories (c) and (d) must be used to formulate questions to be put to social decision based on social aims and values. The decision-making process is thus finally political in nature. Costs and benefits must be recognised as being weighted by political conviction. Open discussion will render the task of the scientists both easier and more difficult. Easier, because less ambiguous and more explicit, more difficult because of the need to render complex issues transparent to non-expert eyes. The essence of scientific beliefs is, or should be, the essence of a liberal view of the world. That is, what is important about political views is as much the way in which they are held as what is held. In current dearth of scientific knowledge of the future and a limit to what can be known there lies a danger which must be avoided at all costs; that is the attachment of spurious certainty to predictions of tenuous truthfulness.

The future of world models

In the first place, it is perhaps worthwhile to enumerate considerations of the modelling process. Firstly, there are those issues relating to choice of problem to be studied using a computer model and the way in which the model can best be integrated into the decision process. Can a single model provide the best solution or should we recognise more explicitly that a model is exclusively the product of particular researchers and their sponsors, therefore it cannot be unbiased, and that each party to a decision should rely on its own model? This is, of course, what happens now. It seems to us to be unrealistic to expect any one model to be universally acceptable.

Secondly, there are questions of the organisation and management of the large multidisciplinary teams needed to construct 'global' models. What seems, from our personal experience, to be needed here is a mutual respect and confidence in each other's ability, and a definite humility towards one's own data and conception of the processes and problems. It may, therefore, be unsatisfactory to bring together teams on a short-term, 'one-off' project, although if there is not a good cross-fertilisation with other researchers through continuous exchange programmes ideas can stultify and progress stagnate.

Thirdly, there is the all-pervading question of data and theory. It is surely in the nature of 'understanding' that these will never be resolved completely, the criterion therefore has to be usefulness. For the moment we try to take the best expert opinions as to theories, data sources and parameter magnitudes and so on. When opinions clash, models can be tested to check whether this affects the results. If indeed it does, then further research is required — there are no short cuts and it is

84

naive to pretend otherwise. If the real life decisions are urgent and cannot await the discovery of better theory we use conventional methods for decision and take our chance as we ever did.

Fourthly, there are questions relating to the validation of models. For the 'physical' part of comprehensive global models presumably tested and accurate theories will eventually emerge. (Even so, unpredictable events will continue to interfere with expectations.) For the 'social' part of a model the same may never be true. Empirical tests of social theories are not accepted as ethical and even if they were, the superb adaptability of the social process might always outwit the theory. This may ultimately set the limit on the possibilities for prediction using mathematical models even if problems of measurement are solved. We can endeavour to use the standard tests; to check individual relationships within a model; to check the overall behaviour of the model over historical periods and so on, but one is left with the uneasy feeling that non-linear complex systems are poorly understood and further research is needed into questions of disaggregation, connectivity and other contributors to mathematical stability.

Lastly, those questions relating to the numerical aspects of computer calculations can be grouped together. These include numerical methods for efficiently handling large sets of simultaneous algebraic and differential equations and the development of simulation languages, of combined continuous-discrete languages and of languages for specific application areas. A point made earlier needs repeating here: there is a danger in treating these complex tools as 'black boxes', robust to all situations and utterly dependable and accurate. They simply are not.

We are rather sceptical of the immediate usefulness of large-scale models for greatly assisting the formation of public policies. It is also apparent that our scepticism does not assume that these models will not eventually be useful but is based rather on pessimism about our general lack of understanding of the world in which we live. But despite this, as we have already shown, when a framework such as a 'world model' exists it does provide pointers as to the information which is most urgently needed to complete that particular world model and in this sense models, even today, are useful aids to efficient research. On the other hand, models can only ever be a partial representation and there is always the danger that important issues will be dimmed in our thinking if we rely too heavily on model results. So there are, to us, two paramount issues to be tackled before these models will realise their full potential. The first is the establishing of a satisfactory, comprehensive and consistent knowledge base. The second is the satisfactory integration of models (both their development and results) into policy-making situations. Of course, what is 'satisfactory' is the big question. We are unconvinced that at the moment present models, despite their formality and precision, approach the standards required.

Even though the present policy-making process has many faults, not least that it is often scarcely democratic, even though it is vague and of course we do not understand how it works, it is fair to say that it does work after a fashion. One reason for this is that it has evolved many checks and balances. What models should eventually provide are not only indications of any further checks and balances required but also a perspective by which the various issues can be judged. At present, models have been used to suggest new important issues although it can be argued that the perspective they bring to bear in the face of other factors is hardly satisfactory.

How can this situation be improved? Presumably by dint of continuing hard work and formulation of a series of realistic goals for each stage in the modelling effort and by a continuing process of the interactive education of both policy makers and model builders. This involves considerations at all stages in the model's development from problem definition to choice of numerical algorithms in the computer application.

However, to counter this 'black view' of models that we have preached in much of this paper we must admit to being slightly schizophrenic. We return therefore to our starting point and move from the short-term future models (the time scale for this pessimism was of the five to twenty year variety) to their eventual potential applications. Here presumably all things are possible and all options are still open. Unfortunately this cannot be true of the real world where inevitably many options have already been lost through lack of foresight or availability of resources, human and material, which might not have been had we had better planning tools. One thing that always impresses about science fiction (especially that which restricts itself to more or less sound physical theories) is that its speculations often come to pass much earlier than the authors ever imagined. Even if models are nothing more than 'science fiction' it may be important that we should use them to seek out worthwhile goals.

References

1. Isaac Asimov, *Foundation* (Panther 10807) and *Empire* (Panther 13555), *Second Foundation* (Panther 17135)
2. Jay W. Forrester, *World Dynamics* (Cambridge, Massachusetts, Wright-Allen Press, 1971)
3. See a collection of papers, 'The Limits to Growth controversy' prepared by the Science Policy Research Unit, University of Sussex, UK, *Futures*, Vol. 5, Nos. 1 and 2, February and April 1973. Also published in *Thinking about the future* (London, Sussex University Press/Chatto and Windus, 1973)
4. M. Mesarovic *et al*, 'An interactive decision stratum for the multilevel world model', *Futures*, Vol. 5, No. 4, August 1973, pages 346-366. Also 'An alternative strategy for the Club of Rome,' *Futures*, Vol. 5, No. 4, August 1973, pages 421–423
5. D. Meadows *et al, The limits to growth* (New York, Universe Books, 1972)
6. This discussion of the Bariloche work is based on a personal visit. The authoritative South American account should be published in early 1974
7. P. van der Grinten and P. de Jong 'Werelddynamics'. *Chemishe Weekblad,* May 1972
8. 'The biggest world model so far', *Nature*, 241, 160 (1973)
9. J. Clark, H. Cole, R. Curnow, M. Hopkins, *Feasibility study of a socio-economic model for Europe*, Report to EEC Commission, May 1973
10. Hermann Hesse, *The glass bead game* (Penguin, 1972)
11. D. Meadows *et al*, A Response to Sussex, *Futures*, Vol.5, No 1, February 1973, pages 135–152

Neither there nor then
A eutopian alternative to the development model of future society

Jim Dator

My problem here is how to try to make a contribution towards a solution of a problem without thereby becoming part of the problem. My topic was assigned to me in these symbols: 'Is the social model of development in East and West necessary and still valid within certain limits for the countries of the South?' I was not absolutely certain what that question might mean, so I rephrased the question as follows: 'Is there a crisis in the image of the future which is signified by the term "development" such that the image has lost much of its appeal to persons in "developed" and "developing" countries alike? If so, why? And if so, what might replace it?'

Assuming my rephrasing of the question to be acceptable, I proceeded to answer the question as I restated it:

Yes, there is a crisis in the concept of 'development' as an image of the future. Both the concept and the fact are being seriously attacked from within the 'developed' nations by environmentalists, counter-culturalists, humanists, and assorted neo-nostalgics. It is also being attacked by some people within the 'developing' countries who allege it to be a debilitating mirage, an imperialist ruse, or both.

Yet I find that many people still defend the notion of development most enthusiastically. Many 'modernisers' abroad, and leaders of the underclass generally, emphatically demand the right to 'develop', and despise environmental, humanistic, or any other attempts to deny them their long-awaited share of what they consider to be the 'good life'. Within the developed countries, too, the 'average man' seems unwilling or unable to turn away from the goals of growth, even when he is told that he has no choice but to do so.

So what is to be done?

I first considered the arguments for and against 'development'. Then I attempted to make a case for — and to ponder the arguments against — a model of future society which I consider to be 'beyond development'. That is to say, I have become convinced that the 'development' model, as I understand it, is not the most suitable one for the future of *any* society of the present. Moreover, I do not find much that is compelling in any of the steady-state, neo-traditional, counter-cultural, or otherwise future-shocked alternative images that I have encountered either. Instead, I believe that it is possible and responsible to imagine and work for the attainment of a future which seeks not the 'limits' to growth, but the 'transformation' of growth; not the recapture of golden pasts, but the invention of more human futures; not the frightened re-conforming to our group and 'the ecology', but the chance for each of us to be more individually freed, interactively loving, and environmentally creative.

If you don't like my image of the future – and you probably will not – then please try to create a better one of your own. Indeed, that is precisely what I hope to provoke you to do by the stunning failure of my own audacious attempt.

I make no claims that the future I imagine will actually come to pass, however. It is an image, a model, a 'dialogue focuser' (to use Robert Theobald's phrase), a generalised social goal, a yearning, an expression of nostalgia. It is not a blueprint. Blueprints and other ironclad plans are the last thing we need, in my opinion. But compelling visions – that is something else.

Not utopia, though – not in the sense of the ideal, impossible dream-world of Nowhere. Not dystopia – the horrifying world of so much science fiction and ecologicised fiction science of the present. But eutopia – an image of a feasible, desirable place (and time) which is significantly different from, and better than, the present or the past.

Development: an American model

'The Country that is more developed industrially only shows to the less developed the image of its own future.' Karl Marx, *Das Kapital*, 1905 English edition, Vol. 1, page 13.

There are a number of models – and actual examples – of 'development' available for any person to choose to be illustrative of a 'development' image of the future which spurs, or can be used to spur, 'developing' nations or groups onward. I have chosen a model which was produced in the USA in the late 1950s and early 1960s. I rely mainly on W.W. Rostow *The Stages of Economic Growth – A Non-Communist Manifesto* (ref.1); Max F. Millikan and Donald L. M. Blackmer, eds, *The Emerging Nations – Their Growth and United States Policy* (ref.2); and Robert L. Heilbroner, *The Great Ascent – the struggle for economic development in our time* (ref.3). Though other works could be utilised, Rostow, Millikan and Blackmer, and to a lesser extent, Heilbroner are sufficient examples of the American development model of about 1960, not only because they were very specific and self-conscious attempts by American scholars to define the process of development and to influence American policy towards the 'developing' countries on the basis of the American theory and model, but also because much American policy apparently was shaped on the basis of this model. The volume cited by Millikan and Blackmer, for example, was originally written in response to 'a request from the [United States] Senate Committee on Foreign Relations for a report on economic, social, and political change in the underdeveloped countries and its implications for United States policy' (Millikan and Blackmer, p.vi).

While I will give some attention later to a Marxist critique of the generalised American model of development, I consider the Marxist alternative model which might be found in the writings of Marx and Engels, in the writings and actions of Lenin, Mao, or Castro, or in Russian, East European, Chinese, Cuban, North Vietnamese or other Marxist countries of the present to be simply an alternative development model which, for my purposes, is more similar to the American model than different from it.

One similarity is expressed in the quotation from Marx himself which heads this section: industrially developed nations generally show an (the?) image of the future to less developed countries. Thus, whether undeveloped nations are attracted to the American, Marxist (or some other) model of development, to the extent that they wish to be developed, at all, and see their role as being one of imitating and 'catching up' with a developed nation, to that extent they have no crisis in their image of the future (such as seems to plague many of the 'developed' nations) and are dismayed and angry with those who suggest that they should not themselves 'develop' along the lines they seek to follow.

As I.L. Horowitz puts it, 'The choice for other nations is not between capitalism and communism – or perhaps between oligopoly and state capitalism – (the growth of the economic "mix" in both societies makes a choice implausible if not entirely impossible) but rather between the industrial ideology shared by Americans and Soviets, and the pre-industrial ideology of the Third World Countries' (Irving Louis Horowitz, *Three Worlds of Development,* 1966; 1972, p.54f).

Thus while the method of achievement and operation, class-base, purpose, and beneficiaries of industrialisation between capitalist and communist countries is different, and importantly so, they are alike in their admiration of industrialisation itself, and equally ruthless (though displaying it differently) with persons and institutions which attempt to resist industrialisation in favour of underdevelopment or non-development.

Both the communist and capitalist models of development have similar manipulative views of man in relation to nature. Both believe that humans and machines can and should be linked purposively by science and politics to achieve human advancement. Both also believe that a great deal of human freedom – or laziness – must be sacrificed or rooted out in order to achieve the 'labour discipline' needed for high productivity. In short, if (as I will argue later) a certain attitude towards 'work' and the human manipulation of labour is necessary in order to achieve an *industrialised* society, then much Marxist ethics and aesthetics (as applied, though differently, in the USSR, China, and Cuba, for example) can be viewed as a functional equivalent of the 'Protestant Ethic' in the same way that the Confucian-based 'Bushido' of the Tokugawa period was a functional equivalent of the Protestant Ethic in Japan.

Goals of development

The goals of development, according to the American model, are very clear indeed: to 'look like' the USA; to 'modernise'; to 'Americanise'; to 'Coca-colonise'; to have (at least) the *material things* which the USA has (at the time the model was formalised).

Horowitz states it clearly: 'A developed society is one made up of the social structures, technologies, and lifestyles that exist today in the First and Second Worlds' (meaning the USA and the USSR, respectively, p.50). The widely respected economist, Paul Samuelson, expresses the same sentiments in reverse by asserting that 'an underdeveloped nation is simply one with real per capita income that is low relative to the present-day per capita incomes of such nations as Canada, the United States, Great Britain, and Western Europe generally.' (7th edition, *Economics* 1967, p.737. Note that no mention is made here of 'The Second World'.)

Robert Heilbroner speaks of 'the great ascent – the struggle for economic development in our time' as 'not merely a struggle against poverty. The process which we call economic development is also, and in the long run primarily, a process through which the social, political, and economic institutions of the future are being shaped for the great majority of mankind' (p.10*). Heilbroner observes that among the persons who are seeking to aid developing nations to achieve development (government officials, university scholars, and foundation experts are some he mentions explicitly), 'a specifically American point of view colours the prospect of the Great Ascent. In part it is visible as a tendency to stress the socially constructive and appealing aspects of development. In part it is apparent in the tacit assumption that the political processes of development are discussable in an American political vocabulary. In part, again, it shows itself in the unthinking premise that rapid economic growth in an undeveloped area will generate the same social and political contentment as it does in America (p.13).

So what is to be done? To answer that question, the model builders had to develop a theory of social development, and this they did. Though there are many different variations of it – some of them quite significant – one of the clearest, and almost certainly one of the most influential statements, is the 'stages of growth' argument of W.W. Rostow.

If the USA (and, to a lesser extent, Western Europe) are also examples of 'developed' nations, then an explanation of how the USA developed would suffice – or would it? That is a major problem. *Must* developing nations follow the American/European path, or can (must?) they find their own way? Indeed, is the American/European experience unique and totally non-replicatable? If the American/European experiences are 'essentially' necessary indicators of successive stages, then what must (should) developing nations repeat, and what can (should) they ignore, by-pass, or do differently? (Similar questions can also be raised about the Marxist model.)

In *The Stages of Growth* Rostow postulates that it is possible to identify all societies, in economic terms, in one of five categories: the traditional society, the preconditions for take-off, the take-off, the drive to maturity, and the age of high mass consumption.

However, beyond 'high mass consumption' Rostow admits it is impossible to predict, but observes that Americans, at least, have behaved in the past decade as if diminishing relative marginal utility sets in, after a point, for durable consumers' goods; and they have chosen, at the margin, larger families. . . . (Rostow, p.11f).

Something, after all, might lie 'beyond development' even for the 'developed' nations! But it is only a hint, and while we will see that a decade of 'diminishing relative marginal utility . . . for durable consumers' goods' has lengthened into a generation, and swelled into a protest against 'development', Americans did not

---

*p.10 refers to the page number in the book by Heilbroner which is cited in full at the end of the chapter. Similar references occur after all quotations.

long follow the peasant's alternative of 'at the margin, larger families'. Indeed, the phenomenal population growth of the 1950s appears to have 'permanently' ended, birth rates are dramatically lowering, and 'zero population growth', now a major rallying cry of a variety of anti-development counter-cultures in the USA, seems an imminent possibility. But Rostow shows little indication that development will not actually continue to be a major goal of the USA and certainly for the underdeveloped nations of the world.

In the famous Appendix B to the second, 1970, edition of *The Stages of Growth,* wherein Rostow answers the critics of his theory, he says that 'On my view there is a good deal that is automatic about the process of growth after take-off, and a good deal that is not' (Rostow, p.173). 'Sustained growth', he says, 'requires the repetition of the take-off process. It requires the organisation around new technology of new and vigorous management; new types of workers; new types of financing and marketing arrangements. It requires struggle against constraints created in the previous generation or two around the peculiar imperatives of an older set of leading sectors now no longer capable of carrying the economy forward at its old pace' (Rostow, p.175). However, 'There is . . . nothing in theory that decrees a society might not, after a certain degree of modernisation, decide to pull back from the effort as a matter of political and social policy' (Rostow, p.176), and business cycles, trend periods, wars, revolutions and 'self-inflicted distortions' (*loc cit*) can retard or reverse development once it has begun. But the overwhelming tendency will be to keep going towards maturity, Rostow believes: 'Men in societies must continue to struggle to keep growth moving forward . . .' (Rostow, p.177).

'It is as impossible to be absolutely precise about the development of a nation as it is to be absolutely precise about the growth and development of a human being, but the general rules, stages, and problems of growth are very well known in both instances: 'In economic growth, as in any form of analytic history', Rostow asserts, 'we are dealing, then, with a biological science' (Rostow, p.179). Though he does not continue the analogy, this may mean that Rostow's theory operates at a genotypic but not a phenotypic level of social analysis.

Having now a general understanding of the 'stages of growth' theory, let's consider briefly what specifically is done in order to achieve 'development.' What must a country do to move from being a traditional to a developed country, according to the American model of development?

Economic development as the base of social development

It should be clear that the key to 'development' — to achieving an 'American standard of living' — is economic growth. Other things may be necessary too — educational, political, religious and social reforms — but a necessary (and perhaps in some cases, even sufficient) condition for development is simple the economic growth.

'Economic development is a necessary condition for the satisfaction of the host of new aspirations that fill the minds of the members of a modernising society', say Millikan and Blackmer (p.43). 'Another reason for the dependence of modernisation on economic change is that economic change is one of the key factors causing the changes in values [which] we associate with the modernisation process' (p.44). This is an especially interesting comment because, as Rostow indicated, and as I will develop in more detail shortly, a 'change in values' on a significant part of the population of a traditional society is usually said to be a *prerequisite* for economic growth to commence in the first place! 'The essential difference' between developed and developing countries, Millikan and Blackmer contend, 'is that in the developed societies, continuing growth of output is a regular and normal feature of the economy, whereas the national per capita product of traditional societies fluctuates erratically around a static norm' (p.45).

Let's consider this point more closely. Millikan and Blackmer pinpoint a major difference between developed and underdeveloped countries as being the continuity or the discontinuity of economic growth. Developed nations, they imply, are adaptive, they continue to grow, they do not 'collapse'. That's how you can tell they are developed and not underdeveloped.

Heilbroner also describes the 'Shackles of Backwardness' (Chapter III) and 'the panorama of misery and desperation visible on the faces of the Arabs, Indians, Africans, and Bolivians who stare at us from the pages of magazines and television screens' and have 'shocked us like a glimpse into the inferno' (Heilbroner, p.11).

The basic problem, according to Heilbroner, is productivity – or, rather, unproductivity. A backward society is unproductive. It is agriculturally unproductive, partly due to the fact that there is typically both landlordism and no primogeniture, but also because of overcrowding, bad farming techniques and soil exhaustion. However, 'the low level of agricultural productivity is largely due to an inability to apply capital to the productive process.'

'And yet this economic handicap is not the ultimate cause of underdevelopment', Heilbroner states. 'For the shortage of capital itself testifies to a still deeper problem. This is the absence of those social attitudes and institutions which create capital' (p.44). Heilbroner then goes on to differentiate the typical backward 'peasant' from the entrepreneurial 'farmer' of the developed West who is 'price and cost conscious, quick to adjust his output to the signals of the marketplace. He [the American farmer] is technically minded and ready to alter his practices if something demonstrably better comes along. He is, for all his traditional ties to the land, very much an economic man' (p.44).

But the problem lies not only with the peasant, Heilbroner observes. It is found throughout all people in a traditionally underdeveloped country. Heilbroner describes the major characteristics of various categories of underdeveloped peoples before concluding that 'One need examine only the history of one of the most advanced underdeveloped nations, such as Mexico, to encounter in living reality the social types we have described: inert peasants, undisciplined workers, cheating businessmen, and rapacious government officials' (p.50).

So how *is* economic growth ever achieved in the face of such overwhelmingly negative odds? Heilbroner is, of course, profoundly pessimistic about the probability of developing countries ever making it to development, but he sees the accumulation of capital as the key issue. This accumulation, he argues, can only result by the society saving – refraining from using all the energy and materials it needs – and the 'freeing of labour and resources from consumption goods production so that they may be applied to capital goods production' (p.74f). Cumulative economic growth will take place, he predicts, provided these savings result in output rising faster than population.

Such is the process of economic growth, as Heilbroner describes it. There seems to be no other way to achieve the 'Great Ascent' to development. Yet, he notes gloomily, 'very few of the underdeveloped nations have reached a point of capital accumulation at which anything like a "take-off" seems possible within the near future. For most of them, the next generation will be one of pre-industrial rather than industrial building, of preparation rather than of cumulative achievement' (p.122). In the meantime, patience, diligence, and hard work are recommended for their morally purgative effects.

Similar stages and processes of economic growth are contained in other sources, such as Millikan and Blackmer.

Let's return to Rostow for a moment to consider the question of economic growth in a slightly different light. He believes that economic growth is 'the consequence of the progressive, efficient absorption into the economy of new technologies', which are 'essentially, uniform at particular times in modern history ...' (Appendix B, p.179f). But he warns against facile generalisation pointing out the need 'to identify the extent to which particular technologies have been absorbed efficiently into the economy, and the sequence in which they are absorbed.' Here Rostow comes face-to-face with a crucial problem (there are others!) in his theory: the types of technology available to developing countries now are incredibly more varied, and imply significantly different socio-environmental consequences, than did those which were available to Europe, the USA and Japan a century or so ago. Doesn't this make a fundamental difference? Doesn't this kill the 'automaticity' of the 'stages' theory? Rostow insists that it does not:

> It is undeniable that the changing pool of technology has caused differences in patterns of growth at different periods. . . . On the other hand, when the sectoral pattern of growth is examined over, say, the past twenty-five years, what is striking is the broad degree of continuity with the past in the sectoral task of growth at different stages in the contemporary world (p.181).
>
> The tasks of the precondition period (e.g., in Black Africa) remain as they have long been: the build-up of infrastructure; the education of a generation of modern men; the creation of institutions which can absorb technology and mobilise capital; the expansion of agriculture to permit the growing cities to be fed; the generation of increased export earning capacity. The first range of industries tends to remain light-consumers' goods, including the most classic of all – factory-manufactured textiles. Beyond take-off there are, evidently, sub-sectors of the engineering, chemical, and electricity industries not known before, say, 1915. But the capital deepening that goes with the drive to technological maturity in, say, contemporary Mexico, Iran, Taiwan, and Turkey bears an *unforced* family resemblance to that which occurred in, say, post-Civil War America, post-1870 Germany, post-1905 Japan and Russia.
>
> In short, the alternation of technologies with the passage of time must be taken into account. It does change the sectoral *content* of the stages of growth for the latecomer (as, indeed, it did to a lesser extent for the pre-1914 latecomers). But it does not make the approach to growth via stages – rooted in a sequence of sectoral complexes – less germane; nor does it repeal the relevance of economic history to the contemporary world (p.182, italics added).

The clear implication is that since the stages have been followed even so recently as

1945 to 1970 then they must be followed in the future:

- There are stages of economic growth which are necessary in order to have social development. The West has gone through them.
- These stages must be followed in order, though more or less rapidly, if one wishes to develop.
- Economic growth requires a Puritan-like attitude towards work and savings. Capital must be acquired from voluntary or forced internal savings, and/or from external gifts – or thefts.

Without pausing to question these conclusions here, let us remember them as being major assumptions of the American development model.

Development in non-economic sectors of 'underdeveloped' nations
While the model clearly emphasises the primacy of economic development as a precondition to all other social development and human improvement, it is also clear that economic growth itself is systemically related to all other aspects of society. Thus, from the 'developers'' point of view, supporting 'economic men' and institutions must either somehow *pre-exist* the economic take-off or they must be created. Unless appropriately supporting attitudes and institutions develop along with the economic institutions, true development will never occur.

While there will be considerable differences of attitude and ranges of institutions among industrial societies on the one hand and among pre- (and post!) industrial societies on the other, on the average, there should be even greater differences in attitudes and institutions between industrial societies and pre-industrial societies than there are among them. Hence, it is very difficult not to 'destroy' everything else in traditional society when you attempt to eliminate poverty by industrial technologies.

Yet, precisely because there are crucially significant differences in the societal impact of 'turnpikes and canals' compared with 'automobiles, satellites, and antibiotics' – and because, in my opinion, we may be rapidly coming into possession of technologies that may have profoundly different societal impacts than 18th or 19th or early 20th century technologies had – the basic 'automaticity' of the 'stages' theory strikes me as being essentially invalid, and, more than this, profoundly dysfunctional and undesirable. In fact, that will be the basic position I will take when I come to my suggestions: there is no reason I find compelling why a nation not already well on the way to 'development' needs to follow the 'stages' or processes of the American (or any other) development model. Indeed, there is every good reason, I shall attempt to show, why 'developing' nations should 'take-off' and fly right on past the 'developed' nations into a post-industrial (post-development) society.

That is one reason why Millikan and Blackmer's little comment about traditional societies *vis-à-vis* developed societies seemed so quaint: 'In the traditional societies, old patterns of behaviour persisted even though new circumstances required changed behaviour, and the society ceased to function well enough to prevent disaster' (p.5). By this classification scheme, 'developed' societies may have become very 'traditional', and traditional societies modern.

This is a good place to make another observation about social adaptability. Millikan and Blackmer observed that 'It is virtually never the social group in control of a traditional society that leads the way to modernisation. That group, which finds the traditional social order satisfactory, virtually always resists change, even if the society is threatened from without and change is necessary to resist that threat' (p.10). If a shift from a developed to a post-developed society is necessary or desirable (as we will see some people arguing shortly), then I wonder if a similar comment about the current elites in developed nations might be in order?

Are certain attitudes prerequisite for development?
We have already seen that our informants about the American model of development have indicated that the questions of values – shall we say, the articulation of human needs? – is crucial to the entire matter of development. Millikan and Blackmer pointed out that to bring about basic economic change, men 'must desire to use their energies in manipulating the physical world rather than regard such activity as demeaning and distasteful' (p.21).

Heilbroner, too, had some judgements on the traditional peasant in contrast to the entrepreneurial farmer of a developed society, and he also criticises the worker's reluctance to accept factory work and the modern economic system.

But can't something be done about this? American workers – like Russian workers, no doubt – must eagerly trundle off to the factory or mine each day. Why do workers in developed countries seem to have so much better attitudes towards factory work than do workers in underdeveloped countries?

This is a problem that has worried many a student of development. Is there a fundamental and primary difference between the attitudes of people in developed and underdeveloped countries such that a major obstacle to becoming developed is the 'underdeveloped' attitude itself? Must the citizens of a developing area possess and be able to act upon a set of attitudes which are more or less specifically appropriate for 'development'?

I regret to say that the answer seems to me to be 'yes'.

Several years ago I examined the preconditions for a nation's economic development. 'It seems that they must include as a minimum the possession of certain levels of natural resources of material and men, the existence of institutions and technology enabling these resources to be advantageously employed, and the wit, will, and chance to use them' (Dator, ref.6). I was mainly interested in the problem of 'will': 'That is, what values seem to support successful economic development?' In seeking an answer, I surveyed the literature on development and conducted a sample survey of nearly 1000 randomly-selected Tokyo citizens. The evidence appeared to me to be overwhelmingly clear:

• There is a set of attitudes, which Max Weber identified as the Protestant Ethic, and which, at least in Europe and the USA, does in fact seem to be related to significant differences between Protestants and non-Protestants, that is positively related to economic development along industrial lines. 'This ethic includes, among other things, the beliefs that it is a positive good for men to work, and that it is better that they do work than that they do not; that science and technology are also good things to be used for the benefit of man; that man, through the application of reason, can, in significant measure, control his environment; and that while religious consciousness is important, religious or superstitious beliefs should not discourage or prevent men from doing well economically and materially in the world, but rather that religion supports — indeed, requires — such endeavours' (Dator, p.24).

- There exists in those developed areas of the world which have not had sufficient contact with Protestant Christianity for it to have been a factor, a set of values which appear to be 'functionally equivalent' to those of the Protestant Ethic. These are especially evident in Japan in the form of the neo-Confucian ethic called the *Bushido*, as they were identified by Robert Bellah, in *Tokugawa Religion*, though a Zen Buddhist forerunner had been pointed out by Hajime Nakamura in the writings of Suzuki Shosan. (See Hajime Nakamura, ref.7, *The Spirit of Capitalism in Japanese Buddhism.*)
- For the sustained, systematic, and rapid economic growth which 'development' implies, if no equivalent to the Protestant ethic exists, or if none can be evoked as with Marxism or, imported, as in Hawaii, then economic development along the industrial lines described by the 'stages' model is probably impossible.

This conclusion may be disputed but if it does stand up, the interesting point to me is not that 'underdeveloped' attitudes are a 'problem' that must be overcome in order to push unwilling people into the glories of the industrial age. That seems to have been *yesterday's* problem. In the future, if the necessity and/or desirability of industrial-style development has somehow come to an end, then should we try to 'de-develop' people's attitudes? What attitudes might take the place of those of the 'Protestant Ethic' and its analogues? How urgent — if at all — is it for people to divest themselves of the 'obsolete' values of development? Must we *force* people to give up their development ethics, as we have forced many persons to acquire them in the first place? Or is growth-through-hard-work a necessary and permanent part of a high standard of living that cannot possibly be avoided?

The problem of 'growth' in the history of economic theory

The problem of growth is fundamental to the entire economic enterprise. Many of the critics of 'development' seem to assume — or pretend to assume — that economists believe growth will go on forever. These critics frequently state their case against development as though they were telling economists something new when they point out that there are limits to growth. Yet whether we turn to Adam Smith or to Karl Marx — or whether we focus our attention on economists before or after their time — we find that the question of how to keep an economy growing occupies a major portion of their thought. Smith's *Wealth of Nations* can be considered a handbook for economic development, and *Das Kapital* certainly does not ignore the issue! Both economic theoreticians dealt with the problem of take-off and maturity, and both were very much concerned with the issue of stagnation.

Moreover, what is frequently posited as 'new' about the crisis in modern economic theory — the relationship between population growth and economic development – is also actually quite old. It was the fear that population growth would overtake labour productivity that caused Thomas Carlyle to dub economics 'the dismal science': No matter how economists tried to figure it, after take-off comes maturity, after maturity comes stagnation (due to population growth and inherent land/resources limitation), and after stagnation comes death. Holding off stagnation – and hence postponing death – is what economics is all about, as far as the classical economists were concerned.

What seems to have prevented the worst fears of the classical economists from being realised – according to most capitalist economic theoreticians – is not some new discovery in economic theory, but rather scientific and technological innovations that have expanded productivity and resources. The question for our time may be: Can innovation be sustained, or have we reached the limits to human creativity?

Problems in the measurement of 'growth'

Defining 'growth' and 'development' verbally is one thing. Determining how to achieve growth, and framing policies for it, is something else. But both the development planner and the politician want to know whether or not economic development is in fact being achieved. How do you measure development? How can we determine whether or not we actually are better off than before as a consequence of our plans and policies?

W.W. Rostow and Simon Kuznets pointed out that 'conventional' economic measures of growth, such as GNP, employment rates, estimates of labour productivity, balance and types of trade, and even per capita investment rates and levels, did not substantiate the 'stages' argument. Rostow replied by saying that his theory was akin to the biological sciences and thus defined general rules of organic growth which could only with difficulty be applied to the precise growth of actual organic economies. Then he criticises aggregate data analysis pointing out that data are the product of bureaucracies collected for 'policy purposes and to answer intellectual questions posed in the past – sometimes in the distant past' (p.187).

Regardless of whether or not this is a sufficient defence of why the data do not seem to support his theory, it is an excellent point. It reminds us that the problem of measurement is extremely important and always related to both theory and politics. Moreover, the things we measure and the ways we measure them tend to get out of date like everything else in our society, and bureaucratic convenience and routine often seem to take precedence over what others think are more urgent data needs. This very much seems to be the case today.

Criticisms of the 'development' model

There is no shortage of criticism of the model of development which I have just presented. Indeed, the very concept of 'development' has *always* been criticised from a number of postures in the USA. Within the American community, 'progress', 'science', 'technology', 'growth', 'reducing human welfare to quantitative indicators', and all the rest have always been attacked. And I feel that the novelty of the criticisms I am about to summarise is more apparent than real and may be more the product of our selectively forgetting the complexity of American history than anything else.

I would like to consider four types of criticisms of the development model. One is a recent Marxist criticism of the model directly as it applies to Latin America, the second is the complaint launched by 'the ecology counter-culture world order' movement. The third is an 'ethical' argument presented by Denis Goulet, while the fourth is the neo-nostalgic dream.

1. *The Bodenheimer 'dependency-imperialism' critique*

Susanne J. Bodenheimer has recently offered a critique of the development model which focuses quite sharply on the assumptions and ideological bases of the American model, both as it has been formulated in American social sciences generally, and as it has been applied by American scholars in the analysis of Latin America specifically (Bodenheimer, refs. 8a and 8b). Bodenheimer not only presents her own penetrating dissection of the development model (as we have displayed it above), but she also summarises some of the other Marxist criticisms of it as well. I offer hers as a Marxist critique of the American developmental model. Whether other Marxists accept it as such I have no idea. I would very much like to know.

The American model of development in general, but especially as it applies to Latin America, is rooted in American epistemological preconceptions, specifically those that derive from the capitalist, class-based nature of American scholarship within American capitalist society. Though it pretends to be value-free and objective, it is nothing more than a thinly veiled cloak to hide the truth of Latin America's centuries-old dependency on a succession of capitalist countries, of which the USA is simply the most obvious among the most recent. The 'underdeveloped' condition of Latin American countries has virtually nothing to do with the 'causal' factors embodied in the American development theory. In fact, in many instances, the development model is empirically totally in error. 'Under-development' is structurally related to and dependent on America's capitalistic 'development'. Indeed, it might not be too much to say that underdevelopment is *caused by* America's capitalistic development, and that many of the things which the development model say should happen in order for Latin America to move along the 'appropriate' stages of growth actually have the effect – if not the intent – of making Latin America *less* developed and more dependent than if these events had not occurred.

The development model of American social science, of which the Rostow theory is one very good example (but not the only good example) has 'four analytically distinct but integrally interrelated themes.' (Bodenheimer ref 8a, p.99; all further quotations in this section are from that article.)

The first theme is the 'cumulative notion of knowledge and development' (pp.100-105). 'On the substantive level, the cumulative concept has appeared in the form of the continuum theory, according to which development and "modernisation" proceed in a continuous, linear progression from "traditionalism" to "modernity" ' (p.102). 'The continuum theory takes the abstract processes of social and economic development as independent variables whose occurrence need not be proved since they often serve as assumptions of the whole model. It leaves room neither for alternative routes or objectives of development, nor for the possibility that the current route may lead to a dead end or to economic stagnation' which Bodenheimer indicates is actually what is happening in Latin America. Indeed, 'in many regions of Latin America, "underdevelopment followed upon and did not precede development" ' (p.103).

Any notion that the current situation in Latin America is essentially like that of America or Europe (or Japan) in the 18th or 19th (or early 20th) centuries is totally erroneous, Bodenheimer implies. 'The very fact the Latin industry is increasingly controlled by foreign capital, for example, suggests that industrialisation cannot have the same bearing upon national development objectives in contemporary Latin America as it did in 19th century Europe' (p.104).

Referring specifically to Rostow, Bodenheimer observes that 'the very presumption that Latin America is at a "lower" stage of the development process already completed by the industrialised nations leaves little room for the possibility that Latin America is experiencing a qualitatively different type of development and that the future of Latin America will not be a replica of the American present' (p.105).

A second theme in the American model of development, according to Bodenheimer's analysis, is that of stability (pp.105-107). 'The prevailing model takes as both desirable and necessary change that does not disrupt the existing order and permits continuity with the past and present' (p.105). Indeed, 'real change' is viewed as being dangerous and pathological because of the basically conservative nature of the structural-functional approach which underlies many forms of the American model.

A third theme in the American model is the now frequently castigated 'end of ideology' fixation of American pluralism (pp.107-110). The American model assumes the fundamental compatibility and goodwill of the developers and the 'developees'. Class antagonism, in the Marxian analysis, is specifically rejected. Interest group politics, based upon a fundamental consensus, is looked for and expected.

Finally, Bodenheimer discusses 'diffusion' as a fourth theme in the American model of development (pp.111-121). 'The diffusion hypothesis contains two major assertions about development: (a) that development occurs largely through the spread of certain cultural patterns and material benefits from the developed to the underdeveloped areas; and (b) that within each underdeveloped nation similar diffusion occurs from the modern to the traditional sectors' (p.111). Yet this is not the case in either instance. Rather, the rich in Latin America have become richer, and they have done so specifically at the expense of the poor whom they have *made* poorer. 'In fact, as a number of studies indicate, the developed sectors have blocked progress in the traditional sectors and have advanced materially only at the expense of and through the exploitation of the latter' (p.112).

Similarly, the gap between the developed countries – primarily America – and the Latin American countries is widening, not narrowing, *because of* (not in spite of) the policies of the developed countries. 'It has been shown in a number of studies ... that foreign investment by the USA and other industrial nations in underdeveloped areas has resulted in a net outflow of capital from the underdeveloped to the developed nations – a decapitalisation of the former' (p.113). And this decapitalisation occurred by *following* the prescriptions of the development model.

'The net effect of Latin America's integration into the world market and economic relations with the advanced nations has been 'structured underdevelopment' and increasing financial dependence, rather than the promotion of autonomous progressive development' (p.114).

Should we even take the American model seriously? Does it reflect anything concrete at all? Is it even valid for the USA, much less 'valid and necessary' for the rest of the world? 'In fact, it must be asked whether [its] concepts . . . are any more accurate as descriptions of the American past than as prescriptions for the Latin [American] present and future. In a rather significant sense, then, the paradigm-surrogate seems less a projection of the American *experience* than of the predominant American *myths* . . . ' (p.121, italics in original).

The basic point Bodenheimer makes is that Latin American underdevelopment – and underdevelopment elsewhere in all probability – has very little to do with such things as 'lack of n-Achievement', 'inadequate domestic capital', 'too small a middle class', and the other things attributed as caused by the development model. It is simply that 'Latin American economic growth is still governed largely by the needs of foreign economies' (p.125). 'Nevertheless, it would be an oversimplification to maintain that the international system causes underdevelopment directly; rather, it does so indirectly, by generating and reinforcing within Latin America an *infrastructure of dependency* (p.126, italics in original).

Moreover, because of this infrastructure, it would not be enough for Latin America somehow to free itself of control by developed nations; it also must 'implement a profound, anti-capitalistic, socialist transformation of its own socio-economic order' (p.127). Hence, 'harmony' can by no means lie in store for the future of Latin America, or for the future of any developing area. The future must hold conflict. There must be a socialist revolution.

Inasmuch as Bodenheimer describes the dependency model and the imperialism model of 'development' as being actually the same theory viewed from the 'inside' and the 'outside' respectively, perhaps the quotations from the 'ideology' model will suffice to indicate the nature of her critique. Only more evidence – which she abundantly provides in both articles – could indicate more precisely than I have already done the bill of particulars which she presents. I refer you to her sources, predominant among them is a plethora of scholars, especially André Gunder Frank, Ivan Illich, and Paul Baran.

2. *Models of anti-development: the ecology 'counter-culture' world order critique*

Whatever might be said for the power of persuasion of the Bodenheimer attack on the American model of development there is still this much affinity between them: they are both models of *development*. Bodenheimer's criticism is essentially no more than that the true 'dependency-imperialism' model of American development *prevents* underdeveloping areas from developing. And that is plainly bad.

But there is an increasingly large — and increasingly vocal — number of persons of various persuasions around the world who doubt the value of development itself. (Reference 9 is a list of the more influential of these.)

The three interrelated problems of environmental pollution, rapid population growth, and abrupt social change have been recognised at least since the early days of the Industrial Revolution. London was aware that it had a smog problem in the 13th century. Malthus posited his famous formula about the deadly imbalance between population growth and food supply in 1798. The Chadwick report in England in 1842 deplored the then current and still used methods of sewage disposal, and recommended treatment which would permit its use as fertiliser.

In the USA, if Carnegie and Rockefeller and the robber barons characterised the 19th century, Ralph Waldo Emerson and Henry Thoreau were also extant and influential. The isolated, agrarian, non-mechanised, rugged individual on the frontier had always been part of the myth of America. Conservation and 'ecology' are by no means recent discoveries. Indeed, the word 'ecology' first was used in the 1870s.

Yet, what I will call 'the ecology movement', as a significant *political* component, is quite new. Rachel Carson's *Silent Spring* might have been a harbinger, and Paul Ehrlich's *The Population Bomb* (1968) might be considered to have been the movement's ideological manifesto though the movement only really got off the ground in the early 1970s. This newer movement claims that until relatively recently, human beings had to take their environment completely for granted — as a given to which *they* had to adapt. But in 'adapting' to nature, humans developed technologies which now are themselves so drastically modifing the environment that it is questionable whether it will long be fit for human habitation. Various catastrophies are likely to happen within the lifetimes of most people now living. Whereas scientists used to assume that humans could look forward to habitation on the planet Earth for millions of years, now we are warned that we may have less than one hundred years — or less than twenty-five.

> In general terms the ecological crisis is the largely unanticipated consequence of four interrelated conditions: the development and accelerating expansion of engineering technology based on pure and applied science; the unprecedented rise of economic productivity per capita in countries where such development has occurred; the unprecedented exponential increase of population, a byproduct of the technological revolution; and the persistence of an age-old assumption, sanctified by classical economic theory, that the wastes and residues of production and consumption can be endlessly discharged into the atmosphere . . . at minimal costs and without significantly injurious consequences for the human population (Margaret and Harold Sprout, ref.12 p.6).

At the basis of these four consequences is a certain attitude towards life which is said to characterise Americans (and perhaps the inhabitants of all developed countries generally). The environmental architect, Ian McHarg (*Design with Nature*, Garden City, N.Y. Natural History Press, 1969), and the historian Lynn White, Jr. blame the Judeo-Christian heritage.

> Christianity, in absolute contrast to ancient paganism and Asia's religions . . . not only established a dualism of man and nature but also insisted that it is God's will that man exploit nature for his proper ends (White, 'The historical roots of our ecologic crisis', *Science,* March 10, 1967, p.1204).

The particular cluster of so-called Judeo-Christian values may actually be more nearly those which McClelland characterised as 'n-Achievement' and which I suggested earlier are analogues to the 'Protestant Ethics' as a necessary precondition for 'development' generally.

Yi-Fu Tuan, however, warns against comparing Eastern theology to Western practice, as White does (Yi-Fu Tuan, ref.13). Non-Judeo-Christian religions permitted and encouraged as much environmental modification as has Christianity, relative to the specific level of technology available. Moreover, during the 'Dark Ages', Christians lived as 'ecologically balanced' lives as do any peoples who are restricted to pre-industrial, agrarian technology:

> Buddhism in China is at least partly responsible for the preservation of trees around temple compounds, islands of green in an otherwise denuded landscape. [These landscapes were denuded by 'unecological' agriculture and hunting practices.] On the other hand, Buddhism introduced into China the idea of cremation of the dead; and from the tenth to the fourteenth centuries cremation was common enough in the southeastern coastal provinces to create a timber shortage there. Large parts of Mongolia have been overgrazed by sheep and goats. The most abused land appeared as sterile rings around the lamaseries, whose princely domains pastured large herds, though the monks were not supposed to consume meat. In Japan, the seventeenth-century official and conservationist Kumazawa Banzan was inclined to put most of the blame for the deforestation of his country on Buddhism; the Buddhists, he contended, were responsible for seven-tenths of the nation's timber consumption. One reason for this grossly disproportionate consumption was that instead of living in grass hermitages they built themselves huge halls and temples.
>
> Another example of fine irony concerns that most civilised of the arts: writing. Soot was needed to make black ink, and soot came from burnt pines. As E. H. Schafer has put it, 'Even before T'ang times, the ancient pines of the mountains of Shan-tung had been reduced to carbon, and now the busy brushes of the vast T'ang bureaucracy were rapidly bringing baldness to the T'a-hang Mountains between Shansi and Hopei.'
>
> Although ancient pines may already have disappeared from Shan-tung by the T'ang dynasty, from the testimony of the Japanese monk Ennin we know that large parts of the peninsula were still well-wooded in the ninth century. The landscapes described by Ennin provide sharp contrast to the dry, bare scenes that characterise so much of Shan-tung in modern times. Shan-tung has many holy places; the province includes the sacred mountain T'ai-shan and the ancient state of Lu, which was the birthplace of Confucius. The numerous shrines and temples have managed to preserve only tiny spots of green amid the brown. Around Chiao-chou Bay in eastern Shan-tung a conspicuous strip of forest lies behind the port of Ch'ing-tao. It is ironic that this patch of green should owe its existence not to native piety but to the conservation-minded Germans.
>
> The unplanned and often careless use of land in China belongs, one hopes, to the past. The Communist government has made an immense effort to control erosion and to reforest. Besides such large projects as shelterbelts along the semiarid edges of the North, forest brigades of the individual communes have planted billions of trees around villages, in cities, along roads and river banks, and on the hillsides. A visitor from New Zealand reported in 1960 that as seen from the air the new growths spread 'a mist of green' over the once bare hills of South China. For those who admire the old culture, it must again seem ironic that the 'mist of green' is no reflection of the traditional virtues of Taoism and Buddhism; on the contrary, it rests on their explicit denial (Tuan p.248).

But in spite of this, most members of the ecology movement do seem to believe that American or western values and institutions lie at the root of the problem and must somehow be changed.

> . . . the primary question is not who is going to have the power, nor is it moral or immoral war, and it is not capitalism vs. socialism. The first question is life! The primary question is

the recognition of an undeniable ecological reality, and its understanding, which dictates a 'radical' political stand. The question is the re-creation of a total environment which encourages life and growth, rather than death and destruction. Quite literally, what we are striving for is the physical and psychic survival of the human species on this planet. Our politics and our economics must be secondary, nevertheless intimately related, to the real issue of life and death. This is the ecology of revolution (Wagner, ref.14a, p.47).

Whilst some of the more 'radical' pronouncements call for decentralised, ecologically-determined regional governments, this is certainly not typical of the movement generally. Indeed, James Ridgeway in his study of the politics of ecology observes that:

Ecology offered liberal-minded people what they had longed for, a safe, rational and above all peaceful way of seeming to remake society, limiting the growth of capitalism, preserving the natural resources through pollution control, developing a more coherent central state; in short establishing programmes and plans for correcting the flaws in what many perceived to be a fundamentally reliable, sound political system (Ridgeway, ref.14b, p.13).

We would agree that Ridgeway's analysis is the more nearly a correct assessment: most ecologists generally call for governmental regulation of business and other human activities – pollution control, population control, technology control, limitations on mobility, and the like. They also generally argue that not only are these issues too much for local governments to solve, but that they are too much for the states and even the national government to handle in most instances. Air (or water) pollution doesn't stop at some political boundary, it is pointed out. We are all one gigantic global system – spaceship Earth – and our problems can only be solved by regulations enacted and enforced at the most expansive level of government – the world.

Two books recently highlighted the perceived need of global action and global government in order to solve our ecological and ecology-related social problems. One of them, *The Limits to Growth*, (ref.15) reports a world model which analyses the global problem and states that only global solutions are adequate to save the planet. But while the 'Commentary' states that,

We affirm finally that any deliberate attempt to reach a rational and enduring state of equilibrium by planned measures, rather than by chance or catastrophe, must ultimately be founded on a basic change of values and goals at individual, national, and world levels (p.195),

the volume nowhere specifies how this change might be accomplished or how the world or nation might be structured after the change has occurred. In fact the authors frankly admit that 'the model cannot tell us how to attain these models' and they steadfastly refuse to generalise beyond the model.

But another volume which appeared at about the same time as *The Limits to Growth* is not so hesitant. This is Richard A. Falk's *This Endangered Planet – Prospects and Proposals for Human Survival* (ref.16). Falk is affiliated with the World Law Fund (now called the Institute for World Order), an organisation of concerned individuals which, through the medium of a worldwide series of World Order Design Seminars, is attempting to develop specific plans for saving the world from ecological and military disaster by the establishment of transnational world government.

Falk's book is a statement of the analysis and probable solutions to the problems of 'this endangered planet'. He feels that the world is threatened by four interrelated problems: the nation-state based war system; the exponentially growing world population; the insufficiency and rapid exhaustion of our global natural resources; and the environmental overload of geometrically expanding pollution.

In Falk's view, these four interrelated dimensions lead to two scenarios for the future: one where little has been done to change the thinking of leading governments, the other where 'the political leaders of the world cooperate to save mankind from the present prospect of annihilation'.

In 'The future as projected from the present' (pp.419-431) Falk sees the blind continuation of present trends: the 1970s are characterised by 'the politics of despair', when 'people will increasingly doubt whether life is worth living' (p.421). This leads to the 1980s and 'the politics of desperation' which 'breeds the attitudes, tactics, and reactions of revolution and counter-revolution' (p.423). The 1990s will witness 'the politics of catastrophe' as 'the mounting pressure on the environment' interacts with desperate and totalitarian attempts by the sovereign states to restore order, resulting only in bringing about 'the downfall of the world system as we have known it' (p.429) in the 21st Century, 'an era of annihilation'.

But Falk does see hope for us in preventing that scenario from becoming reality, *if* we act courageously and rapidly in the present. His positive scenario is based on a change of political consciousness among the leaders of the world, which enables the leaders to transcend their petty sovereign state concerns and seek – and achieve – global solutions to the socio-environmental crises facing us now. In the 1970s which is 'a decade of awareness':

> The point is to convince as many people as possible that we are in the midst of an emergency, that the traditional priorities, aims, and conflicts need to be subordinated, and that there is a way out, but it involves change, sacrifice, and danger for all societies (p.433).

Once this awareness has been reached – and reaching it is not suggested as being an easy task – then the 1980s can be 'the decade of mobilisation' in which transnational – rather than the currently narrow national – elites will play an important role. The actual change to world order will occur in 'the decade of transformation' of the 1990s where 'a new kind of political ethos will have enlarged the idea of politics to include man-and-nature, as well as man-and-society' (p.434) leading us, by the 21st century, to 'the era of world harmony':

> The applications of science and other branches of knowledge will be guided by service to the maintenance of harmony on the planet. A less homocentric scale of values will underlie political decisions. The welfare of plants, animals, and machines will all be considered benevolent in this kind of humanism. It is a humanism only because the whole process is conceived of and worked out by man, as if man were hired as an architect to rehabilitate the ecosphere inhabited by all that exists on earth (p.436).

These two drastically different scenarios by Falk are not, in a very important sense, different at all. They both seek – and expect, and perhaps even prefer – centralised, uniform, bureaucratic solutions; one ineffectively, the other effectively perhaps. The world order scenario would simply attempt to do on a grander scale what cannot be done, it is argued, by the nation states separately. Structures are not radically changed – just enlarged. It is a change manufactured by intellectualised elites, not popularly based. The only changed consciousness (in Reich's meaning) deals with ecological – not socio-political – matters. Thus consider Ridgeway again:

> The ecologists argue in radical, indeed revolutionary, terms for the re-organisation of society, development of a new political economy which would eliminate ruinous competition, limited production, restructuring of society along more geographical lines, and in general a social order which could enable men to better accommodate themselves to the planet.
>
> Yet, despite the revolutionary rhetoric, [they] relate more easily to the politics of liberal reform. Beneath the revolutionary rhetoric are arguments for policies which would lead to a more efficiently managed central state, a benign form of capitalism . . . (Ridgeway, ref.14b, p.194f).

Although I do not endorse everything that Ridgeway says elsewhere in his book, my analysis of 'action ecology' handbooks – such as Ehrlich's *The Population Bomb* (ref.11), Garrett DeBell's *The Voter's Guide to Environmental Politics,* John Mitchell and Constance Stallings' *Ecotactics: The Sierra Club Handbook for Environment Activists,* and Sam Love's *Earth Tool Kit* (ref.17) – leads me to the same conclusion as his on this point.

There are exceptions in the literature of ecological politics to this type of reformism and simple pressure group formation. To begin with, from the left, some traditional socialists and Marxists have tried to tie ecology to their brand of Marxism, eg Gus Hall, *Ecology: Can We Survive Under Capitalism?* (ref.18). New Leftists, too, early got on the bandwagon. One such example is *The Earth Belongs to the People – Ecology and Power,* written by R. Giuseppi Slater *et al* (ref.19). This thin volume links over-population and pollution with capitalism, and asserts that these things could be solved simply by overthrowing the system and accepting socialism.

Somewhat related to – and yet significantly separate from – the ecology movement is something I call 'counter-culture consciousness':

> There is a revolution coming. It will not be like revolutions of the past. It will originate with the individual and with culture, and it will change the political structure only as its final act. It will not require violence to succeed, and it cannot be successfully resisted by violence. It is now spreading with amazing rapidity, and already our laws, institutions and social structure are changing in consequence. It promises a higher reason, a more human community, and a new and liberated individual. Its ultimate creation will be a new and enduring wholeness and beauty . . . a renewed relationship of man to himself, to other men, to society, to nature, and to the land . . . is the revolution of the new generation (Reich, ref.20, p.4).
>
> Fortunately, there is no need to discuss ways of initiating change, since change is already in motion . . . our task is to optimise the transition from one pattern of culture dominance to the other (Slater, ref.21, p.120).
>
> If the resistance of the counter-culture fails, I think there will be nothing in store for us but what anti-utopians like Huxley and Orwell have forecast (Roszak, ref.22, p.xiif).

There is no beauty in development in the eyes of Roszak, Slater, and Reich. Roszak says that citizens of developed nations live in a 'technocratic society'. Reich writes similarly of the 'corporate state', and Slater believes that 'American culture at the breaking point' is trying to 'kill anything that moves'.

Against the falsifications of the fully 'developed' state a new consciousness, a counter-culture, is said to be rising. The members of the counter-culture, according to Roszak are 'only a strict minority of the young and a handful of their adult mentors' (p.xii), a diverse group as he describes them: 'To one side, there is the mind-blown bohemianism of the beats and hippies; to the other, the hard-headed political activism of the student New Left' (p.56).

These divergent individuals are said to form a more or less coherent cultural unity. They are differentiated from the supporters of the corporate, technocratic state by their common consciousness.

Reich describes three levels of social consciousness which are somewhat reminiscent of the three types of social personalities postulated by David Riesman in his influential book, *The Lonely Crowd*. Titled simply 'Consciousness I', 'Consciousness II', and 'Consciousness III':

> One was formed in the nineteenth century, the second in the first half of this century, the third is just emerging. Consciousness I is the traditional outlook of the American farmer, small businessman, and worker who is trying to get ahead. Consciousness II represents the values of an organisational society. Consciousness III is the new generation (p.16).

In the language of Rostow's theory of development, Consciousness I may be the last vestige of the traditional, pre-industrial or industrialising society; Consciousness II embraces the values of a mature, developed, industrial society; while Consciousness III may be the emerging values appropriate for a post-scarcity, cybernetic society (which, of course, Rostow did not envision).

It is Consciousness III that mainly interests us here because of Reich's insistence that it is leading to a non-violent, but nonetheless revolutionary, change within America and probably all developed societies.

> The new generation, by experimenting with action at the level of consciousness, has shown the way to the one method of change that will work in today's post-industrial society: changing consciousness. It is only by change in individual lives that we can seize power from the State . . .
> All that is needed to bring about change is to capture its controls – and they are held by nobody. It is not a case for revolution. It is a case for filling a void, for supplying a mind where none exists. No political revolution is possible in the United States right now, but no such revolution is needed (p.304f).

Slater adds the point that with the disintegration of the corporate state and the rise of the counter-culture, 'the diffusion of power could occur with little change in the *formal* machinery of government, which, after all can lend itself to a wide range of political types' (p.146). Indeed, unlike the members of the ecology movement, neither Reich, Roszak, nor Slater suggest an *organisational* change in government. What is needed, one might say, is a *spiritual* change – a religious conversion to a new consciousness. Roszak quotes Norman O. Brown with approval:

> The next generation needs to be told that the real fight is not the political fight, but to put an end to politics. From politics to poetry . . . Poetry, art, imagination, the creative spirit is life itself; the real revolutionary power to change the world . . . (p.118).

These are revolutionaries of a totally non-Marxian variety, as they themselves insist.

Indeed, they are anti-Marxists from the point of view of Consciousness III, because Marxism itself is at Consciousness II, they believe. 'Nothing could be more old-culture than a traditional Marxist,' observes Slater (p.97). Roszak is even more emphatic. He spends an entire chapter comparing the position of Norman O. Brown to Herbert Marcuse on this point, and spreads aspersions to Marxism throughout the rest of his book. For example, Roszak admires the attitude of French students during the May 1968 uprisings who declared that if they were Marxists, then it was of the Groucho variety (p.4) and whose slogan was that 'a revolution that expects you to sacrifice yourself for it is one of daddy's revolutions' (p.48). For themselves, they sought a dialectic of liberation. 'From this viewpoint it becomes abundantly clear that the revolution which will free us from alienation must be primarily therapeutic in character and not merely institutional' (p.97).

Perhaps I should try to spell out what 'counter-culture consciousness' is said to be since it is this consciousness – and not a detailed political programme or proposal for world reform – that seems. to embody the present and future political significance of the counter-culture.

According to Reich, the last 200 years of revolutionary science and technology have modified social and human relations continuously and irrevocably (p.351). In an attempt to deal with this rapid change, we have developed several inauthentic consciousnesses and are now struggling to give birth to an appropriate one.

Consciousness I and II have, in varying ways, both made humans subject to their environment and their technology. 'Consciousness III is an attempt to gain transcendence' (p.351).

Thus, Consciousness III is essentially a matter of a new life style. It is an attempt to adjust – or, more correctly, adjust anew – to nature and technology. It is not opposed to honest, enhancing work – but it is opposed to current role-defined and alienated jobs. It is not opposed to morality – but it refuses to accept impersonal, externally defined moral standards. It is not opposed to truly participatory democratic decision making – but it exposes and declines to be a part of the fraud of bureaucratic and corrupt politics. It does not reject education – but it does deride the robotising indoctrination into being a role-playing cog in the Corporate State that is called 'education' today. It is not opposed to reason – but it does rebel against the cruel and inhuman logic of 'crackpot realism' which prevents us from rejoicing in our bodies, our senses, our eroticism through music, personal dress, and dance.

Consciousness III rejects all extra-personal authority, maintaining that only a personally determined moral standard has validity and honesty. It insists that the need for competitive, zero-sum personalities is at an end: 'Given an abundance of material goods, the possibilities of a human community are finally made real, for it is now possible to believe in the goodness of man' (p.354).

While there are similarities among our three informants on the counter-culture, they also have their decided differences. One is the important place accorded individualism by Roszak and Reich in contrast to the anti-individualism of Slater. The second is the belief by Reich that technology may be used to free humans in ways previously impossible. Roszak, in contrast, is violently opposed to modern technology, and Slater observes that 'one of the major goals of technology in

America is to 'free' us from the necessity of relating to, submitting to, depending upon, or controlling other people. Unfortunately, the more we have succeeded in doing this, the more we have felt disconnected, bored, lonely, unprotected, unnecessary, and unsafe' (p.26).

3. An 'ethical' critique of development

In my thought and reading for this discussion, no book or article impressed me as having more relevance to the topic assigned me than Denis Goulet's *The Cruel Choice – A New Concept in the Theory of Development* (ref.23). That is to say, *if* I could believe in the paradigm of what passes for a 'realistic position' today, then the most encompassing, scholarly, challenging, and *humane* diagnosis and suggestions for reform that I have encountered are continued in his book.

The 'cruel choice' (actually choices) in the title involves decisions we must make in seeking answers to two basic *ethical* questions which form the content of the book: 'What *kind* of development is human? And *how* must such development be obtained?' (p.330). This book, unlike many I reviewed, is concerned at least as much with ethical questions as with technical ones, and, indeed, weaves the two together superbly. Moreover, it is a book written more compassionately and less self-righteously than, say, Ivan Illich in his way, or Suzanne Bodenheimer in her way, would have handled the same material. Goulet is less certain he understands all the answers, though I believe he well comprehends the nature of most of the questions.

Goulet does not accept for a moment a Rostow-type 'development' model. 'Underdevelopment', he states clearly and repeatedly, 'is not merely the lack of development or a time lag in achieving industrial strength, productive agriculture, or universal schooling. Underdevelopment is an historical by-product of "development" ' (p.38). One of the major problems he addresses squarely in this volume is how to sever those relationships between developed and developing countries which result in continuing underdevelopment. Another problem that receives his steady attention is the relationship between the traditional values of pre-developed societies and the values implied in the developmental process. He is not romantic about the values of underdeveloped societies, but neither does he believe that those of developed societies are superior and hence must prevail over traditional views.

Thus, he seeks to find those goals of development that 'relate to fundamental human needs capable of finding expression in all cultural matrices at all times' (p.87), and thinks he has found such fundamental needs and values: life-sustenance, esteem and freedom.

> Development is a particular constellation of means for obtaining a better life. Irrespective of possible other purposes, development has for all groups at least the following objectives:
> - to provide more and better life-sustaining goods to members of societies;
> - to create or improve material conditions of life related in some way to a perceived need for esteem; and
> - to free men from servitudes (to nature, to ignorance, to other men, to institutions, to beliefs) considered oppressive. The aim here may be to release men from the bondage of these servitudes and/or to heighten their opportunities for self-actualisation, however conceived (p.94).

In discussing the 'dialectic' nature of development, (by which he means, in effect, what one learns from the experience of trying to actualise one's goals) Goulet remarks, 'Some countries may experience just enough development to destroy their traditional values irrevocably, but not enough to become modern except in a socially pathological manner' (p.107). Indeed, the counter-cultural critics of the very idea of development would probably argue that this precisely defines the characteristics of all modern societies!

Goulet believes that the *goals* for development can be meaningfully achieved *only* if the ethical principles – 'that all men must have enough in order to be human, that universal solidarity must be created, and that the populace must have the greatest possible voice in decisions affecting its destiny' – are followed. He makes a fundamental point that is all too easily lost both by persons favouring and those opposing development in general, or particular actions taken in the name of development:

All progress is fundamentallly ambiguous . . . No one knows whether development, once achieved, can make men happier or reduce conflict among them. But development can eliminate certain well-known forms of human misery (p.120).

There is no false romanticism in Goulet's estimation of value of poverty in a traditional society but neither does he falsely glamourise material possessions:

The fullness of good and the abundance of goods are not synonymous: a man may *have* much and *be* mediocre or *have* little and *be* rich. Nevertheless, men need to have a certain quantity of goods in order to be fully human. The corollary is that some goods enhance man's being more than others. These statements cannot be understood unless we first inquire into the reasons why men need to 'have' goods at all (p.128f).

The reason why men need to 'have enough in order to be more' is simply that 'all organisms must go outside themselves to be perfected'. 'The ontological significance of needs resides in this: that if man were fully perfect, he would not need to need' (p.129). It is in this sense, then, that having goods makes humans better. Goods 'perfect' (I would rather he had said 'complete' or 'enchance') humans – they enable them to be more human (or less! Such is the fundamental ambiguity of progress he referred to above and this is also why the means as well as goods of 'development' need to be reconsidered continually through the 'dialectic' experience of time.).

I am far less convinced that Goulet is correct in emphasising 'universal solidarity' as a second major principle. I presume that he is attempting to find a basis for minimising conflict among humans, and emphasising cooperation, and if so, then this is certainly good and necessary. It is also a prime focus of the 'World Order' advocates we surveyed earlier, and his conclusions are similar.

However, Goulet does not rise to a very good defence of the principle himself. Unlike the very thorough – and persuasive – case he makes for 'having enough to be more' he merely asserts that 'the "ontological roots" of solidarity are common humanity, mutual occupancy of one planet, and identicial destiny' (p.143), which is not a sufficient basis to convince me of anything. It scarcely distinguishes humans from their relation with all other organisms, and thus seems to reduce itself to an unhelpful truism. Moreover, my basic fear is that 'universal solidarity' will be used – as it so often has before – to beat into submission those persons who see things differently from the person who is empowered to articulate what 'universal solidarity' means in reality. My predispositions — as you will see — encourage me to run the risk of chaos rather than suffer enforced conformity, and hence I shy away from 'solidarity' as a basic principle.

Goulet himself says that 'neither individualism nor collectivism is a sound principle of development. The first renders justice, the second integral human growth, impossible,' but I fail to see a sufficient distinction between 'collectivism' and 'solidarity' to understand how individual human growth can be guaranteed if 'solidarity' is one of three basic criteria for development.

I most heartily agree with his elaboration on the third basic principle, however: 'broad popular participation in decisions.' Not only does he enunciate the principle clearly, but he accepts what seems to be the fact that many people — at their present states of development, if I may use that term — apparently do not *want* to participate, and indeed are truly frightened of the responsibility. They seem to be so accustomed to taking orders and not considering their own interests apart from those of a group or a leader, that atomistic, individualised 'participation' is tremendously alienating and undesirable. This is a most uncomfortable situation which Goulet faces, and does not solve any better than you or I can. To leave it to 'natural' elites who act in the 'best' interest of others does not impress me as the best solution, though it may be the easiest and most common. But neither is it desirable to 'freak out' people by overburdening them with decisions they do not want to make. Rather — as I have tried to make clear in many other contexts — it seems to me that there must be a constant review of the relationship between those decisions which individuals wish to make for themselves because they have a direct, personal effect, and those decisions which they prefer to have made for them by their accepted leaders. Although it may be necessary to have elite rule at certain stages of development, I believe that elites should rule only with a guilty conscience, and 'revolution', which Thomas Jefferson correctly understood to be the 'manure' in which individual human development flowers, should be built into the 'constitution' of all political relationships so that individuals are constantly challenged to make increasingly more of their own decisions, while their 'leaders' make less and less.

Questions — especially ethical questions — about the relationship between 'development planners' and persons undergoing development are excellently considered by Goulet in several chapters.

Let me pause to consider what Goulet means by 'existence rationality', the phrase he uses to describe the basic ontological and epistemological premises of any community. These premises are more or less different, and it is this difference which explains why the conventional designation of 'traditional, transitional, and modern' fails to capture the essence of the distinction between cultures, and hence why 'development' occurs so differently in different places:

> The key to understanding why receptivity varies is 'existence rationality', defined as the *process* by which a society devises a conscious strategy for obtaining its goals, given its ability to process information and the constraints weighing upon it (p.188).

Unless the idea of 'development' is adapted to fit the 'existence rationality' of the group intended to undergo development, it will fail, he feels. From this perspective, it is the idea of what is 'development' that must be changed to meet the conditions of the society, not the values of the society that must be altered to suit the developer's idea of 'development',

> If this is so, traditional value structures do not of themselves create obstacles to change. On the contrary, it is insensitive, narrow impact strategies for inducing change which snuff out traditional societies' latent potential for change (p.192).
> The implications for development planning are enormous: it means that the major task may be to teach planners to appreciate old values instead of educating underdeveloped men to the merits of new plans (p.208).

This would seem to be a repudiation of my earlier insistence that something like a 'Protestant Ethic' was necessary for development to occur. Yet, if you glance back at the conditions I set forth earlier, I believe you will see that this is not the case. Moreover, Goulet's consideration of development here implies that people are being developed against their will. It is a strategy for 'developing' people painlessly, and perhaps by fraud. This brings us right back to the 'participatory' and 'elite' question again, and I can only repeat that I personally am reluctant to force people to develop who do not want to, yet I am even more reluctant to be romantic about the joys of the primitive life!

After a review of existing statements about – and actual attempts to induce – development today, Goulet concludes that most 'development' plans are actually 'anti-development'. They *prevent* rather than encourage development because 'they do not contribute to increased life-sustenance or esteem for the neediest masses, nor do they expand freedom from constraints. Rather, they reinforce domination and create new servitudes' (p.234).

Goulet wishes to find a way around the trap of 'anti-development' into which all current 'development' plans have fallen (he tentatively excepts Tanzania and China, though he professes relative ignorance of how these two nations are actually progressing towards the goals of development they – much less he – desire).

As a first step in discovering 'authentic development', Goulet posits a 'theory of needs'. He concludes that there are three levels of needs: *first order needs* – food, clothing, shelter; *enhancement needs* – self expression; and *luxury needs.*

On the basis of these basic human needs, Goulet postulates three lines of development:

> First . . . . to aim at abolishing misery, not at obtaining affluence. Second, that large-scale austerity must be practised in advanced countries as well as in pre-modern ones. Third, that cultural diversity needs to be actively pursued as a counterweight to the powerful forces of standardisation inherent in modernisation processes (p.252).

I will only comment at this point that I disagree with the first two points on the grounds that they are needlessly conservative, and that I fully favour cultural diversity. Indeed, I would expand it to *individual* diversity. However, I believe that a major problem of the future for which I prefer to plan will not be how to retain diversity against homogenising influences (that was the old problem of industrial society) but rather how to retain a *sense* of individual – or group – *identity* in a situation of much greater diversity than most people encounter at the present.

Rather belatedly in the book – and, I expect as is the case with many others, apparently as an afterthought rather than an integral part of his argument – Goulet reveals that he adheres to the ecological, world order persuasion (Chapter 12, 'World Resources and Priority Needs'). In order to exercise the ecological control that he feels is needed on resource use and allocation, a 'world plan' should be adopted. On this point he is especially sympathetic to the ideas of Jan Tinbergen and François Perroux (284ff).

Finally, Goulet considers *how* to achieve development – incrementally or through revolution. He ends up profoundly pessimistic about either method. We do not have time to be incremental, and, as he sees it, all revolutionary plans are sufficiently flawed as to make them unsuitable. Thus, we are faced with 'the cruel choice' in the title of the book because 'development processes are both cruel and necessary' (p.326).

4. *Neo-nostalgia*

Persons who comment on the possibility of the 'developing' nations becoming developed, increasingly point out that the 'developed' nations must de-develop. And they seem to assume that this is simply too much to expect. I am not so sure. Indeed, the reason why I want to consider some of the components of neo-nostalgia is to suggest that it may provide a very convenient set of 'trends' which will enable the trick of 'de-development' to be turned far easier than we might have thought otherwise.

Erich Jantsch observed that:

> We shall have to get the Western countries not only to prevent their economies from growing, but . . . the West will have to take several steps backwards, lowering the national standard of living of the population, cutting back on consumption, and taking a share in a more equitable distribution of the world's resources. This is going to be very hard on the Western governments, and even harder on the people (ref.24, p.109).

It might be easier than he thought at least as far as the USA is concerned. In the flush of the hippie movement and the 'Movement' movement, a good many commentators – including people like Roszak and Reich – saw the end of one period and the beginning of another. John Lukacs wrote a book entitled *The Passing of the Modern Age* (ref.25) on the subject. He was writing primarily of the USA. But since the USA, with only 6% of the world population, consumes some 35% of world energy and is expected to use even more, it would be a major contribution to closing the gap of inequality between the developing and the developed countries if the USA's level of development were lowered. So I will concentrate on the USA and leave it up to others to decide whether her 'future' is unique. Lukacs is only one who provides us with material out of which we might weave a scenario of such a return, for he points out that 'millions of people came to America straight from medieval parts and portions of Europe; they became American before (and, sometimes, without) becoming modern, enlightened and bourgeois' (Lukacs, p.183). Lukacs uses these three terms – 'modern, enlightened and bourgeois' in a strict historical sense. 'Modernity' the 'Enlightenment' and 'embourgeoisment' are changes in consciousness that have (allegedly) occurred in Europe, but not in America, Lukacs maintains. Americans are dualistically split between pragmatic progressivism and medieval fundamentalism. The medieval aspect seems about to be taking over, but it is the Dark Ages, not the High Late Middle Ages: 'Many things, especially after 1960, showed a reversion to barbarism. What were reappearing were some of the tendencies and habits of the Dark Ages, not the institutions of the Middle Ages. Tribal gang warfare, widespread nomadism, the abandoning of the cities, sexual anarchy, drug culture, the meaninglessness of letters, Dragons, Demons, finally the incipient breakdown not only of law and of order but of many of the comforts and services of material civilisation' (p.188). Lukacs adds in a footnote: 'In the anti-Puritan reaction of British and Scandinavian, Dutch and American youth there is something wild and racial in inspiration: the long lank hair, the Amazonic character of women, the emphasis on physical handsomeness and on a kind of hirsute virility reflects a tendency for something that is far from being 'Leftist', intellectual, international; it is, rather, tribal, barbarian, youthful, narrow-minded and proud' (p.189).

Let's pretend for a moment that we don't know that all he (and Reich and Roszak and others) saw in the late 1960s has not come to pass, but, rather, seems to have ended. Let's pretend that we don't know that all currently available evidence in the USA indicates that the communal and ecological has won out over the tribal and proud. Let's forget that Herman Kahn's *Things to Come* (ref.26) fairly gloats in the apparent triumph of the 1950s over the 1960s.

We can forget these things because the *consequence* predicted still seems likely even though the *style* is much different.

It is, in Lukacs' view, a triumph of unconscious historicity over intended ahistoricism. At the same time, intended ahistoricism has made the past *new*. Unconscious historicity has made the ways and artifacts of the past no longer oppressive, but appealing and novel. The 'novelty of the old' is not confined to youth, nor, actually is the appeal of the past restricted to those who are ignorant of history!

Kenneth Boulding has recently said:

> There is quite high probability that the great age of change is approaching an end, that the peak period of change was my grandfather's life, say, from 1860 to 1920. Today's most probable image of the future is not one of accelerating technical change, but rather one in which the gains of the last hundred years are consolidated and in which there is only a slow increase in productivity – at least in the rich countries – in which there is some catching up on the part of the poor countries, in which the international system becomes relatively stable, and in which [in the developed countries] the world of one hundred years from now looks surprisingly like the world of today, except for a greater awareness of the limitations of the earth and the exhaustion of resources (Boulding, ref.27, p.352f).

In a related, but slightly different vein, Andrew Shonfield argued:

> Indeed it is true that the rate of change has accelerated greatly, that there are technical possibilities whose future social and psychological consequences are extremely difficult to foresee, and they might be disastrous. On the other hand, I think I'm impressed when I look back over the enormous rate of change that we've had in the last quarter of a century, by how much has *not* changed fundamentally (Shonfield, ref.28, p.187)
> Therefore, I think that in our investigations of the future we ought to identify those things which seem remarkably constant (p.188).

Indeed, is technologically-induced change really occurring all that fast? Have we not been selective in the past in our presentation of evidence to show that the pace of change has increased? In the mid-1960s Eli Ginzberg presented a graph which apparently indicated that 'the interval between discovery and application in the physical sciences' of inventions has been consistently decreasing. This chart – and its basic premise – has been widely repeated (for example, John McHale, *The Future of the Future*, p.60ff). But this is disputed in a study prepared for the National Science Foundation (ref.29, p.238), and a study conducted by the Battelle Columbus Laboratories, concludes that the lag time (and hence potential for social impact) has not shown a uniform decline at all. Indeed, the amount of time between invention and discovery was found to be almost totally dependent upon a 'technical entrepreneur' – 'one person who champions a particular scientific or technical activity' (*loc cit*).

The inference is that technologies do *not* have lives of their own. Their diffusion must be entrepreneured, and if there are economic or political factors which might discourage such entrepreneurship, then the lag can be lengthened indefinitely. The technological imperative – the frequent allegation that 'whatever is technologically possible will be done, regardless of whether it is socially needed or not' – thus may not be true at all. Rather *decisions* concerning what research shall be supported, and what innovations shall be diffused, seem to remain the most important factor.

If we pause and look back over the things that have been alleged to cause *both* social and ecological disruption, we discover that the same things are blamed: too much education (primarily in the natural and social sciences – not enough in the classics, basics, and humanities), research (too much 'irrelevant' basic research; applied research not sufficiently dedicated to or guided by genuine social needs), and technology.

What a simple way to get things back under control! Stop the money to education and research and you have pulled the plug on the engine of change! You have satisfied both the conservatives and the conservationists.

It seems to me that there is more than coincidence involved in America's renewed concern with pollution and conservation. It is its continuing search for purity and cleanliness. Here, again, the aims of the young ecologists coincide with the older 'average American' businessman and politician. It is quite true that Americans like to work – indeed, feel they need to work – and that they thus do resent the made-work that occupies so much of the time of modern humans.

The Noble Savage, rooting proudly in the soil, has always been an appealing, beckoning figure. How wonderful if we could discover that the savage, primitive nobility we seek does not lie in foreign lands, but in ourselves, in our own past, in the truths of our traditions that we have been encouraged to forget while lusting after the plastic Whores of Babylon! (whom we despise mainly because they are *plastic*. 'Organic' whores we believe we can live with.)

Can we make anything out of this array of neo-nostalgia? Perhaps not. Perhaps it is only a temporary phenomenon – like everything else in 'developed' society – only temporary and skin-deep.

But wouldn't it make an interesting scenario: Over the next 30 years, America becomes increasingly isolationist, increasingly conservative, less and less interested in – or supportive of – education and research, more and more involved with 'old' agricultural life-styles, technologies, and products.

In the meantime, the developing countries are at last able to move ahead free from either the ideal American model of development, or the 'true' American model of dependency and imperialism.

Is this too much to dare to hope for? Is it too much to even try to *encourage* America to 'drop out' and drop back to the earth – while others soar ahead? I think it is not too much. Indeed, it is so much the mood of the moment – America's new Utopian dream – that it might be interesting to contemplate what such a future might be like. Imagine:

America, Europe, and – except for Japan who is not ready to 'drop out' yet – the developed countries generally pull back from further spectacular growth. They improve their agricultural productivity – mainly by returning more *people* to the soil. They continue some basic research but at a much lower level, not in order to develop or diffuse it themselves, but to turn it over to the developing countries, who will skip over the heavy industry phase of the old development model, and move ahead with the most advanced energy and materials production that the newest in evolving technology somewhat more slowly provides. The developed nations continue to withdraw their religious, military, economic, technical, and educational 'missionaries' from the developing areas. They end foreign aid which has been defined as 'using the money of the poor people in rich countries to pay the rich people in poor countries for their allegiance to the rich people in the rich countries' (*Trans-action*, February 1971, p.63). Indeed, as the currently 'rich' countries retreat into pastoralism and the currently 'poor' countries move into post-industrial societies, adapting the best of the newest 'cybernetic' technologies and thus retaining the best of their older cultures, 'developed' and 'underdeveloped' roles became reversed: America struggles to move from its 'future-shocked' agrarian retreat into the cybernetic age. And Asia, Africa, and South America try – vainly? – to help.

Towards transformation

At one time, Marx could say that the developed nations showed the future to the underdeveloped and Joseph Gusfield made Marx's point even more explicit recently:

> In talking about economic development in the new nations of Asia, Africa, and Latin America, we will suggest that the image of modernity provides just such a utopian description of the future – a description which has the transforming effect which we see as the consequences of utopian ideas in many forms . . .
>
> There is a very clear conception of a possible future which gives direction and issue to current actions. The present, as it exists in the [developed] world, is also a future for others in the same world. It is this that gives a utopian character to politics in new nations today. It is often around this character of the future as a possibility that a great many movements occur . . . (ref.30, pp.76, 81f).

But as we have seen repeatedly in our survey, the developed societies are no longer able to serve so obviously and appealingly as models of the future. Among the members of the developed countries themselves, there is a tremendous wave of pessimism, both about their own futures, and about the validity and necessity of the past they have trodden to the present. There is similar uncertainty among persons interested in developing the underdeveloped areas. What is to be done?

I believe that *one* task of those people who presume to labour in the vineyard marked 'futures research' is to generate plausible images of a better future — eutopias*.

But I do not consider eutopias to be 'blueprints for the future'. I would much prefer that we sail rudderless into the void than that we attempt — or succeed — to peg down all of the world to come (see for example, ref.31, p.346f). Since many people live and as far as I can determine many more will continue to live in 'the shifting and dissolving movement of society', then shifting and dissolving images — and flexible and adaptable people — should characterise our dreams, and not our fears. If I could imagine a real future society of 'no further significant social change' within, say, the next 100 years, then I might feel differently, but, in truth, I can not.

This is my basic point of departure, and if I am in error, then I need to be corrected starting from this point. Do not mistake me. I have no essential quarrel with continuity, stability, and order. I freely admit that most people's image of the future — perhaps for the past three million years — was essentially that nothing much changed: 'Tomorrow will be like today, because today is like yesterday'. Such an image was good, and proper — and essentially correct. As far as the individual and his tribe could see, change did not occur; or if it did, it occurred 'randomly' — out of human control — or cyclically. (Figure 1 depicts such a traditional image of the future.)

With the approach and fruition of the industrial revolution came the first significant break in this image. People who became accustomed to living in industrial societies came to believe in 'progress' because they came to experience continual change, and to reflect on it. They acquired the notion — from their experience and hopes — that tomorrow will *not* be like today, because they recognised that today was different from yesterday. I label this image of the future 'developmental' (Figure 2).

It is the image, you will recognise, that has come under severe attack recently — though, as I have tried to indicate previously, most humans in the world retain traditional experiences, and hence images, and many who did adopt the developmental view (for example, immigrants, or some revolutionaries, or founders of religious communes), may have expected things 'to settle down' in a few generations. They were 'making sacrifices for their children'. Nonetheless, many people — especially many powerful people — came to believe (or act) as if development would go on forever. It is this future of 'straight line growth' that is under attack today. Recognising that exponential lines cannot in fact go on forever, they point out that there must be 'limits to growth'; that the Golden Age of growth

---

*I call such images of a better but possible world 'eutopian' in order to distinguish them from 'utopian' images. The Greek word 'topia' means 'place', and the prefixes 'u', 'eu', and 'dys' designate 'no', 'good', and 'bad' respectively. Hence, a 'utopia' is literally an impossible dream of 'no where,' while I use 'eutopia' to indicate plans and dreams for a much better, but possible future. 'Dystopia' I use to label people's fears and nightmares about tomorrow. Obviously these are not clear-cut distinctions. Perhaps my eutopia is your idea of a dystopia. And who is to say what is forever 'impossible'? I admit it is a subjective classification, but I have found it helpful pedagogically and heuristically.

Images of the future

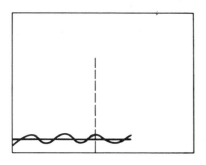

Fig 1 Traditional

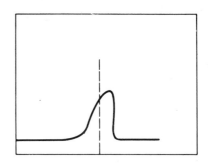

Fig 4 Ecological/traditional

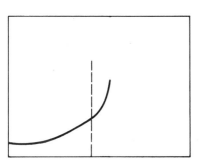

Fig 2 Developmental

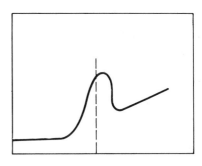

Fig 5 Ecological/steady

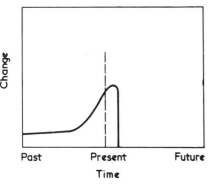

Fig 3 Apocalyptical

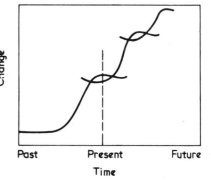

Fig 6 Transformational

and development is over; that — well, here the images differ.

Some say that the system is about to collapse (Figure 3); some say that after a 'time of troubles' we must return to the traditional ways (Figure 4); and others seek a relative 'steady state' which will stabilise at a higher level of technology and culture than that of traditional society, but (at least for quite a while) at a much lower level than the present (Figure 5). The image of collapse I call 'apocalyptical' or dystopic. The last two images are 'ecological'; one 'ecological/traditional' and the other 'ecological/steady'. Although I do not like the two ecological images, personally, I could live with them and I find many of their arguments persuasive. Thus I consider them to be 'eutopian' images generally.

The developmental view *per se* is 'utopian'; impossible and dangerously misleading to promote, at least in its 'pure' form. I must make it clear that I believe there are other 'utopian' images which are not 'developmental'. The 'neo-nostalgia' image is utopian in my view. Traditional, developmental, or ecological images are not eutopic, utopic, or dystopic as such. It depends entirely on what is possible and desirable at the moment — environmentally, biologically, technologically, and culturally.

My own view of the future (it certainly is not 'mine' alone, though my mentors may wish to disclaim my version of it!) I call 'transformational'. That is, I believe we may be entering — and I hope we will in fact enter — a world of continued and increased change. My image may be depicted as a series of overlapping 'S' curves of 'development' (Figure 6) which are intended to indicate that I recognise one aspect of 'the limits to growth' argument, but I find also that frequently when we reach the limits in one mode, a transformation occurs which pushes the old limits aside and places new ones far beyond our reach — for a time.

It is possible that we will reach, or have reached the absolute outer limits beyond which there is nowhere else to go but back. It is also possible that we have moved and are moving faster than most people can handle, and that we must slow down and go back for their sake and for the sake of our environment. It is possible that the 'traditional' image is in fact 'normal' and that all others are pathological.

Yet I seriously doubt that those possibilities are inevitabilities, and I think it is both our failure of nerve and a failure of imagination, coupled with a fad of despair, that makes us think we have no other choice but to return.

I believe that under certain circumstances which I will specify, we can handle much more change than we are currently experiencing — biologically, psychologically, culturally, and environmentally — and that we can and should work towards the creation of a world which maximises our ability to live even more humanly in a situation of rapid change. I believe that to do so is much more 'natural' than to try to stop or slow down. John McHale has pointed out that 'evolution is permanent revolution' (ref.32, p.62). I believe that this is certainly so. There is no absolute 'balance of nature' for us to preserve, no necessary and permanent symbiotic relationship between humans and their environment, or between any organism and its environment. The world and its inhabitants apparently have been in 'mutually programming harmony' from the beginning — or forever.

Humans have been fooled by our (to us, long) past experience and environments of relative stability into thinking that change was abnormal and undesirable —

pathological, in fact. To change either the environment or the social structure was taboo because it was dangerous. To preserve and continue the old adaptive responses had high survival probability in a situation where the environment *was* stable and where we could actually modify the environment neither extensively nor predictably.

Recently, and it seems only for a brief while, some of us were lulled into believing that we could change part of the system and retain continuity in the rest. We thought we could make 'amendments' to the constitution of life; that we could patch up things through a bit of legislation here, a bit of land-fill there. What change there was, was to be 'for the better'. Outside of this, nothing was to change — the 'eternal' verities (that is, the adaptive responses of the past), the values, needs and institutions of earlier times were to prevail.

A great many people around the world still believe in development. But a great many more who do not, feel that our only alternative is to go back to the old ways, or, where they exist, to preserve the old ways from development. The old 'revolution of rising expectations' and the newer 'revolution in rising pollution and population' has destroyed the image of development for many people who now fear that further change will kill us all; that our dream of a 'better' world was, in fact, only a dream; that our sin, our hubris, our vain presumption that 'mere humans' can ever do anything right has led us into deep trouble again. Our ancestors, our gods, and Mother Nature (who we are reminded, bats last, and best, and with the biggest stick) are punishing us for our transgressions. And, like well-socialised sinners, we *love* it. 'Beat me, Mother, for I have sinned.'

I disagree. And I do so not only because I just personally enjoy the exhilarating uncertainty of rapid change of this kind, but also because I believe we actually have no other *real* choice.

Denis Goulet wondered how we could avoid the 'cruel choice' of revolution or incrementalism (he was sufficiently impressed with the reality of the misery of underdevelopment to reject 'no change'). Yet I believe there in fact is no choice here. There is only revolution. Revolutionary change seems to me to be unavoidable. I can see no way to avoid continued radical discontinuity. Thus, for us to argue about *whether* we should be revolutionary or not is to miss the point, I believe. Our discussions should concern towards what *goal* and by what *means* — and perhaps to some extent, *how swiftly* — we wish to move.

- To favour bombing, shooting, sabotage, to destroy buildings and people; to tear down the old institutions and power structures — is to be revolutionary;
- to support or encourage or permit the continuation of the current business and political interests of the developed nations is to be revolutionary — and is to support massive structural violence against the underclasses and underdeveloped groups at home and abroad;
- to favour further scientific and technological development is to be revolutionary because such development has been and is the major cause of most socio-environmental change;
- to support retreat, de-development, return to the old ways, slowing down and stopping is also revolutionary in its implications of a radical change from the

institutions and values and unfulfilled dreams of many persons alive now;
- finally, to live in a world where all this — and more — is going on simultaneously is certainly to live in a revolutionary situation, and to act or not to act at all is to contribute to the revolution.

I suggest you try to enjoy the ride. While certainly not all change is beneficial, any attempts at achieving permanent stability are, in my opinion, suicide.

Our experiences — our dreams — of mobility; the kaleidoscopic personal experiences which more and more people (though by no means all people everywhere in the world) are having; the new information which piles in upon our old assumptions; the increasing genetic variability of our former relatively distinct breeding groups all seem to me to be forces which are pushing us towards greater change and towards solutions to problems that do not lie wholly within the traditions of any extant society, nor in their history and myths.

> Man augments [the] diversification [of the earth] by altering the physical characteristics of the land, changing the distribution of living things, and adding human order and fantasy to the ecological determinism of nature (René Dubos, ref.33).

Or, as the *Last Whole Earth Catalogue* impiously, but accurately, said: 'We are as gods, and we might as well get good at it.' Perhaps you would prefer a more humble formula: 'We are all aborigines in a new world.'

In the following image of a 'transformational society', I will attempt very directly to describe how I would *like* the world to be. I am not predicting or forecasting the future. I am making a verbal model of the future that I would prefer.*

A theoretical prelude

A fundamental assumption which underlies this section is that of the reciprocal and feedback relationship between (1) those things which at any point in time are identified as 'human needs' and (2) certain human characteristics and experiences over time. That is to say, from my point of view, human 'needs' are not in any important sense uniform across cultures or time. 'Needs' change. Moreover, there appear to be a wide variety of potential and actual ways in which any set of 'needs' at any point in time might be satisfied. In the past, humans have both identified their needs and attempted to satisfy them in an even wider variety of ways.

To be precise about how needs have been and might be identified and satisfied requires a complete theory of social structure and social change. I will not present such a theory but I will indicate what I consider to be its basic components.

A theory which seeks to explain how human needs are differentially identified and satisfied over time would have to describe the composition and dynamic interrelationship of the following components on both an individual and an interpersonal level:

*An earlier version of this section was prepared for a Colloquium on Designs for Global Futures, conducted by Professor George Kent of the Department of Political Science of the University of Hawaii. My class on Advanced Futuristics, Spring 1972, also worked on portions of the statements on basic human rights and responsibilities. I wish to thank Thomas Mandel especially in this regard.

- biological (especially genetic and physiological characteristics of an individual or group);
- cultural (significant symbols and epistemologies; for example, language, religion, art, play);
- social (regular patterns of human interaction, such as familial, sexual, educational, economic, military, political, to identify them according to one current structural perception);
- ecological (humans' relation to the physical geography, climate and weather, nutrition, architecture, other organic species);
- technological (humans' relations to artificial extensions of human physical and intellectual capabilities; especially human relationships at different 'levels' of technology).

From my point of view, all of these components are operationally interrelated into a dynamic system which has the nature of any system such that 'a change in any component of the system causes changes throughout the entire system'. Hence, determining the true 'cause' of change in a dynamic system is extremely difficult, if not absolutely impossible, without first discovering the legendary 'unmoved mover'.

Nonetheless, I find it helpful to emphasise those changes in human systems and sub-systems which are caused by changes in the genetic and the technological components of the system. To repeat, all of the components mentioned above are mutually interrelated into a feedback system (of more or less precise articulation, of course) so that my stressing the causal relationship of any one component or sub-component on any other might be misleading. However, for analytic purposes I will consider that for human systems, changes in culture, society, and ecology are caused by changes in (1) the distribution of human genetic characteristics and (2) levels of technology.

Quite obviously, the cultural, social, ecological, and technological components (eg, religious structures, marriage rules, nutritional patterns, and medical and birth control technologies) themselves affect the genetic structure. And while McLuhan's famous phrase, 'we shape our tools and thereafter our tools shape us' suggests that we concentrate upon the ways in which tools shape human behaviour and attitudes, nonetheless what tools are developed and specifically how they shape us itself depends largely on the state of the rest of the components of the human system at the time (and preceding the time) under consideration.

I hesitate to labour this point, but if I don't I will probably be accused of being a 'technological determinist' by someone who forgets these caveats. The problem of 'what causes what' in a dynamic system is such that any sort of 'determinism' is sticky business indeed. John McHale, in commenting on his emphasis on the 'organic nature of technology' in a portion of *The Future of the Future* (ref.32) observes that it is not his intention 'to pose some technological determinism as accounting for all human development and change. One could as easily suggest poetry as the determinant, and with as much validity. Rather, the purpose is to re-emphasise the integral nature of human activities, whether labelled technological, religious, economic, cultural, or whatever' (p.95).

The Transformational Society: its values, goals, and human rights

I desire a society* where every human being is wholly free in every way from unwanted control by any other person individually, or by society as a whole or in any of its parts. The dominant unit of my society of the future is the individual person. As a rule, in situations when an individual conflicts with a group of whatever size or significance, preference will be given to the expression of the freedom of the individual over that of the group, though as I will show, this may not always be the actual case.

The primary function of society, and therefore of 'government' at any level in this future, then, should be no more than to provide an environment in which each individual may become self-fulfilled, self-reliant, and self-determined without unwanted interpersonal or environmental constraints. As a rule, each individual should be treated the way that individual wishes to be treated.

Thus, I take this as the criterion by which we shall judge all social and political structures and decisions in this future: Is the individual enabled by them to do what that individual wants to do? If not, then my basic preference has been violated, and every attempt should be made to modify the political structure or decision in order to bring it in accord with the basic value preference.

Nonetheless, such a condition of individual freedom might be meaningless, if not positively threatening and harmful, for individuals without the following conditions also applying:

1. *Every individual has an absolute and irrevocable right to food, shelter, clothing, medical and dental care, and any other material goods and services which are available to anyone else.* In principle, these things should be made available to each individual person's own level of satisfaction. That is, the distributive principle is that of R. Buckminster Fuller's 'bare maximum' rather than current notions of 'minimum standard of living', or 'minimum guaranteed annual income', and the like. If it turns out to be impossible for each individual to have all the 'things' that the individual wants, then each individual should be guaranteed a real 'functional equivalent' to that individual's self-determined satisfaction. Finally, if there are items of true and permanent scarcity which cannot be distributed by the principles above, then these things must be permanently owned by no one, but rather made equally available to all by lot and by rotation of use for fixed, brief periods.

2. *Every individual must have the unlimited right to be physically mobile or to remain in one place.* I do not mean by this that any individual has the right to intrude seriously upon another; each individual also has the right to be left alone from unwanted interference by any other person. Any actual conflict in these two rights may be resolved in many instances by adding that while each person has the right to be left alone, no person is thereby *obliged* to move on – or not to move on. While one person's right of mobility ends, as in the analogous classical situation, where the other person's body begins, in situations of conflict, accommodation and tolerance are first to be sought. However, if two or more people nonetheless wish to occupy the same space or to move in the

---

*By 'society', I mean here any number of individuals who interact physically or symbolically.

same direction, a solution of 'functional equivalence' should next be sought. Only when this is finally determined to be impossible should the conflict be settled by lot and rotation.

3. *Each individual has the right to acquire as much knowledge, information, education and experience about any topic as that individual wishes.* But no one must be forced to learn or receive communication about anything at any time. However, no information about anything shall be denied to anyone who wishes it. This implies an end to many of our recently acquired notions of privacy to the extent that no one can deny another person any information he has that directly relates to that other individual, and all persons should be discouraged from withholding any information from any other person. Still in the final analysis, a person can refuse to tell someone something of a strictly personal nature that has no significant relation to the well-being of the other. All other communication must be open to anyone who wishes to receive it.

4. *Every individual has a right to effective participation in all social decisions which affect that individual.* As far as possible, society should be arranged so that all individuals can act as they please. When collective decisions cannot be avoided, the size of the groups involved in and affected by the decisions being made should be kept as small as possible so that all persons substantially affected by a decision are able to participate effectively in making it. Genuine effort must be made to maximise the efficacy of the individual's effect in decision making, both in process and in outcome. In keeping with my basic rule, I believe that an individual should fully understand and positively consent to any collective decision which negatively affects that individual. If the individual cannot consent, and if a collective decision is made which an individual judges negatively affects his freedom, then he has the right to receive compensation from the collectivity in such manner as to redress the infringement.

5. *Every individual has the right to seek and receive affection,* however and wherever expressed, from any other individual or individuals who wish to display such affection. While any individual has the right to seek affection, any other individual has the right to deny it to that individual, without either person thus necessarily implying offence or rejection.

6. *Any individual has the right to join with any other individual or group* of individuals who wish that person to join with them for the pursuit of any mutually-shared purpose which does not seriously affect any individual not in the group or consenting to its activities. While an individual may accept the consequences of consensus or majority, or any other collective decision-making rule, each individual also possesses the absolute right to resign from the group or to fail to abide by its decision at any time.

To re-emphasise, I believe that each individual should have the right to self-determination and self-expression according to his own criteria, coupled with the real attainment of the objects or goals which he has chosen. It is not enough merely to salute such principles as 'equal access to positions of power' or 'equal economic opportunity' and the like. It is far more appropriate I believe to encourage and enable individuals to choose the level of goods and services, or the amount of interpersonal involvement, (or none) which each person desires, and then to facilitate the fulfilment of these desires. This principle is the guiding key for all the rights previously noted: the society which is defined by these values and rights must be real, desirable, and attainable for each individual, and not be simply the idealisation of an egocentric set of philosophical goals.

The boundaries of the Transformational Society are not complete without including a minimal set of *social* and *environmental responsibilities:*

A. No individual is free to threaten, harm, or kill any other person against the other person's will. (See Appendix for a discussion of the causes of crime)
B. No individual is free wantonly to modify the physical environment beyond his — or other humans' — ability to restore the environment to its previous state; or to a functional equivalent thereof according to the then current limits of technology.
C. Every human being must be made aware of the probable significant effects, both on other individuals and on the ecosphere, of any major or significantly unusual individual and collective action before he acts.
D. Nothing in these three items is meant to prevent a person from doing with himself what he will, nor to limit the varieties of interactions between consenting individuals, nor to encourage us to become too narrow in our conception of what a 'natural' or desirable environment is. But there may have to be some limits to behaviour in these arenas, if this society is to become and remain a viable environment where the satisfied individual is the focus of life.

Some objections to the values and goals of the Transformational Society

There no doubt are a great many reasons why the Transformational Society is either impossible or undesirable. Let me suggest some objections that come to my mind.

1. It is impossible because it is contrary to all human history. Never before has humanity ever seen such a world and therefore there must be good reason why it should not come into existence. If it hasn't been already, then it can't be now, and for some good reason.
2. It is impossible because it is contrary to human nature. People can't stand that much freedom. They demand and need order, not freedom. The Transformational Society expects too much from mere mortals.
3. It is undesirable because it is too atomistic; too individualised. Most importantly, it is what you might expect from a spoiled upper middle class pseudo-intellectual white American in the latter quarter of the 20th century. It is just too ethnocentric for words — or for generalisation to the world as a whole, especially the underdeveloped parts! People don't want to be alone in the way the Transformational Society suggests. They want *community*. To be together, unified by a joint purpose and will.
4. Even if such a society were possible, and even if some people really do want it, it is just too disorderly and violent for most people to tolerate.
5. It is impossible because its basic premise — a society of material abundance — is

impossible. Since I say that the Transformational Society cannot exist unless each person is able to be as economically and psychologically secure as he wishes to be, then it cannot exist!

6. It is impossible because I forget 'X'. My item 'A' conflicts logically with item 'D'. What about that? How do I handle problem 'F'?

Let me try to offer a brief rejoinder to these objections.

1. *The Transformational Society is contrary to history.* I suppose I will have to admit that the society I wish to have in the future is contrary to history, if by 'history' we mean the past five thousand years of recorded human experience, plus the five thousand years before that time which have later been memorialised in myths. But unwritten human experience is much longer than a mere 10 000 years, and there are what seem to me to be good reasons to believe that the past 10 000 years are (1) very poor indicators of what life might have been like for the earlier hundreds of thousands of years of human life, and — much more importantly — (2) thoroughly erroneous as an indication of what life might (or can) be like in the future (see for example Glasser ref.34).

While the Transformational Society may be contrary to 'history', it is not thus invalid because there seem to have been other 'ahistorical' human conditions as well. Indeed, the bulk of human experience may be different from what we have enjoyed/endured and memorialised during the past 10 000 years. Thus, while 'history' obviously means something in forecasting or even designing the future, it does not mean everything. It is especially helpful to us in our design efforts if we can go beyond history to a more useful past which might have elements — especially genetically-imprinted predispositions — which can be institutionally supported to our future advantage.

There are elements of relativity in all human experience. Humans have revealed many kinds of environmental and interpersonal modes in the past. Perhaps that reminder will embolden us to try to make a radical break from 'history' — if that seems desirable — in designing for the future.

It would be pleasant indeed if it were true as some have argued — or if our assuming so helped us behave as though it were true — that we humans have 'cooperation' genetically built into us. If 'conflict' is 'unnatural' — or at least not our basic predisposition — and if 'cooperation' is the more likely human response when freed from the 'unnatural' restraints of civilisation, then we should be greatly encouraged in an attempt to design a world which exploits freedom without fear of unleashing unrestrained and unrestrainable violence.

Finally, if we can indeed organise societies to provide 'abundance' for all, then people can be expected to cooperate — rather than conflict — because they *enjoy* being with other people, and not because they are forced to interact with others either for the sake of their own survival, or because institutions and powerful people require them to.

2. *The Transformational Society is contrary to human nature.* In part, my argument against this objection is the same as the one I made above. Ortega y Gasset said that 'man has no "nature". What he has is a history.' That is, what we

term 'human nature' is nothing but the remembered chance experiences of a human's past.

I believe that the human experience over time has been one of monotonic movement from virtually complete determinism to ever-increasing (though still certainly not complete) 'freedom'. What I mean more specifically is this: earliest humans were (1) genetically similar to other members of their tribe; (2) experientially starved by consequence of their relative lack of mobility (as individuals, not necessarily as a species); (3) intellectually stunted because of inadequate tools for recording, remembering, transmitting, and analysing information about themselves and their environment; and (4) over-socialised into the taboos of their fellow humans.

Most humans in industrial societies are now quite different from earlier humans on all four of these measures, and humans will be hopefully even more significantly different in the future. That is, I believe that humans: (1) are becoming increasingly differentiated genetically – and hence predisposed to be different behaviourally – from other humans who happen to be physically near them; (2) are increasingly 'forced to be free' and forced to make their own untraditional decisions in more and more areas as a consequence of the unique, technologically induced experiences which they increasingly have and for which tradition provides no adequate guide; (3) as a consequence of increased science, education, and information dissemination, have a great many sources of ideas and opinions, and bases for analysis, which may dispose them to feel and act differently from each other and (4) are increasingly forced to live independently of most people who are physically near them – ie, whereas in a tribal (or otherwise 'primitive') situation, *most* of the people one met daily were personally influential in the direction of your life, now most of the people one meets are total strangers, and even those influential in some areas of your life are 'powerless' in others.

Hence, I conclude, whatever may have been 'natural' for humans over the long course of their past experiences is less and less likely to be 'natural' in the future – or the present.

3. *The Transformational Society is too atomistic – and too ethnocentric.* I will have to admit that it may be that my preferences are too ethnocentric, too much the expression of the values of a comfortable American white boy. I have had that charge laid against some of my ideas before, and it may be true. But even if it is true, it still does not render my position defenceless. It may be that it is not ethnocentric in the sense of being values that only Americans (etc) espouse, but in the sense that for a number of reasons, Americans are among the first people in the world to experience what is likely to become a universal situation. To the extent that the longing for transformation is based upon certain types of technology, and to the extent that the USA early possessed and felt the impact of these technologies, then the Transformational Society is ethnocentric. But if other people experience or are likely to experience a technological environment similar to that of the USA (and if changing technology does in fact help change behaviour) then we can expect other people to feel some of the same things Americans allegedly feel, and perhaps to wish to behave similarly as well.

While some non-Americans (in Europe and Japan, of course, but 'modernisers' throughout the 'developing world' also), seem bent on moving into the 'American' present and beyond, it may be that rising ethnical consciousness globally may merge with another 'global trend' — anti-technology — to make the future I prefer totally impossible. It is easy to envisage one scenario for this: If we can be convinced that 'technology' causes 'human alienation' at the personal level, and that 'technology' is invariably used by powerholders to wreak domestic and foreign colonialism at the national level, and that 'technology' causes pollution and resource exhaustion at the global level, then, for God's sake, let's turn off the machine. It's much more 'ethnical' and 'organic' to starve to death by old methods than to try to create and operate new and more humane technologies and institutions.

Indeed, I would agree that this is one choice that needs to be made now: if you want to freeze your present or unthaw your past, then you must turn off the major engine of socio-environmental change which is, indeed, scientifically informed technology. But if you wish to use technology to seek cures to *any* socio-environmental problems, then you had better assess, and plan for, the systemic consequences which even the most meagre technological advance tends to have. You also will have to establish clear criteria for developing some technologies and not others. The matter must be faced frankly, whatever the final choice is to be.

4. *The Transformational Society is too violent and disorderly at best.* If I am correct in assuming that technologically induced experiences with 'freedom' create an expectation and demand for more 'freedom' (as well as helping create 'future shock', 'alienation' and attempts to escape from freedom, which it probably does at one and the same time), then neither the preference for, nor the fact of 'law and order' can result unless you either halt further technologically induced change, or put down the demands for freedom with brute force.

Much of our dislike of 'freedom' is because many of us have been socialised into feelings of inadequacy, actually live lives of such tenuous dependency, and are so uncertain of where the next meal — or next lover — might come from if we lose the one we have, that in any real contest, we tend to favour order over freedom (see for example ref.35). I hypothesise that much (not all) of our fear of chaos will be 'cured' once we know that we are totally secure economically and psychologically. If one meal — or lover — vanishes, there are plenty more around which are easily available to us.

5. *We do not have enough abundance to achieve a transformational society.* I do not need to recapitulate the 'ecology' argument. However, 'facts' and 'trends' are tricky things. I don't believe we know enough about the 'slope of the curves' to be able to *predict* what the future development of many of our objects of interest and concern *will* be. Aside from not knowing the characteristics of the system in sufficient detail to be able to predict its future state, I believe we have tremendous data problems: we do not have very accurate measures of things we need to know in the present, and we have no data whatsoever of many of these things in the past. For this reason, too, 'prediction' is tricky.

Moreover, the 'trends' themselves may be misleading, because the 'limit to growth' in one area need not mean the 'limit to growth' in all areas. What frequently (not always!) happens, in my understanding of the phenomenon of growth, is not an 'end' to growth at all, but a 'transformation' of growth: the end to growth in one mode and its continuation in another. This is the very conceptual basis of my Transformational Society.

Of all things, what we really seem to be running out of is imagination and determination. It is because of the belief that our problems are human, institutional, and technically solvable that I would like to turn my attention to designing the world the way I would like it to be and working for the realisation of this end, rather than spending most of my time trying to figure out what is wrong with the present — and what might be wrong in the future — and who is to blame for it.

Shapes and shadows of the Transformational Society

Can we create a world where humans will behave freely, independently, yet non-violently? At the outset in the design of such a world, we are faced with the problem which Theodore Roszak calls 'the inevitable confrontation between Freud and Marx': Do we try to shape people's consciousness or their environment? Is our main task one of influencing the way people perceive reality, or do we change that reality so as to influence people's perceptions? Does that which is intractable in present life lie within each human organism individually, or is it within the structure of relations and interrelations which humans have (purposely or non-purposely) created?

My argument throughout this study has been that an individual's behaviour is caused partly by his biogenetic structure, partly by the memory and fact of his past behaviour, and partly by his perception and the fact of his present environment. Thus, while we can influence an individual's behaviour by changing his attitudes (perceptions or memory) we are likely to be even more effective if we can alter the actual environment and/or the biogenetic factors which underlie his behaviour.

Hence, I believe that we may be able to utilise two new capabilities in our futures design project: namely (1) our growing ability to pre-influence human behaviour through genetic engineering, chemical alteration, or electronic implantation (or similar and successive techniques); and (2) our developing ability to shape human behaviour through the application of Skinnerian — or post Skinnerian — operant conditioning techniques (See J. Platt, ref.36).

Without question, in neither instance are these tools capable of doing now all that we need them to do in order to help us design and guarantee a Transformational Society. But I believe that they offer us more hope than any other techniques I know about, and I suspect that their strengths and weaknesses are more likely to be revealed as they are used in the service of individual human freedom than in some other likely domains of utilisation or neglect — such as 'National Defence' or 'Crime Control'.

I also will assume that the more optimistic claims about the development of bio-cybernetic technology (meaning primarily the computer and its progressively microminiaturised and organically-grown supporting and successive technologies) can be realised, and the technology utilised as anticipated. Namely:

• that it will be (is?) technologically possible to operate a world where less than

5% – perhaps even fewer – of the potential human labour force is involved in the production and distribution of *all* goods and services;

- that the current clear trend towards wholly synthesised foods will have been completed so that all nutritional and aesthetic human requirements can be met entirely in principle, and largely in fact, by these;

- that we will develop artificial intelligence which is functionally equivalent to human intelligence within the next 30 to 50 years, and that shortly after the first such capability is attained, we will be able to utilise this non-human intelligence in any area where we now use human intelligence – and in new areas now hidden from human perception;

- that our understanding of the operation and management of large-scale human and environmental. systems – and our understanding of the operation and control of ourselves both as individuals and in our socio-environmental context – will increase;

- that through something like 'technology assessment' we will be able to reverse the so-called 'technological imperative' which seems now to mean that we develop and diffuse a technology for some manifest purpose without first anticipating and evaluating its latent or 'secondary' human and environmental consequences. Instead, we should be able not only to develop institutions and technologies to help solve the problems which humans wish to have solved but also to be more mindful of the major secondary effects of those institutions and technologies; and

- that we will have sufficiently reliable 'human and environmental indicators' so that we can personally and mutually evaluate our progress over time.

In this instance, too, we are more likely to come to possess these capabilities at all, and to be able to use them for human freedom rather than human enslavement, if we cause them to be developed for these purposes in the first place and not wait for freedom to 'trickle down' from the profits of military, corporate, or bureaucratic technologies.

Arthur Clarke asks a pertinent question:

'In an automated world run by machines, what is the optimum human population?' (ref.37).

Most people today seem to agree that optimal population – and population growth – are crucially important, first-order questions for any design of the future. But if we are given, not the 'facts' of our world today, but the goals and assumptions of the Transformational Society of tomorrow, what *is* the optimum population? Clarke says:

> There are many equations in which one of the possible answers is zero; mathematicians call this a trivial solution. If zero is the solution in this case, the matter is very far from trivial, at least from our self-centred viewpoint. But that it could – and probably will – be very low seems certain. That is to say, in a fully automated world, why have any humans at all? I can think of no good reason other than habit and self-interest. That will have to do for now. So if the number of humans is to be more than zero, how many? One? Two? My family? My nation? As many as now?

Since my goal is a world of free, and freely-interacting though non-violent, individuals, and since I am assuming that such a world will not be effective *in toto* until bio-cybernated technology has been more fully developed and diffused, I must face the problem of population determination and control.

It seems highly likely to me that biomedical technology will and should be asked to *permit* the complete separation of sexual intercourse from human reproduction, with all reproduction being a highly deterministic consequence of specific human choices. In the Transformational Society, reproduction is not likely to be a consequence of individuals' activities *qua* individuals. Neither should we expect that the 'parents', of a child produced by a conscious bio-medical process should be responsible for the rearing and care of that child. As with reproduction, this, too, will probably be handled 'professionally' and/or cyborganically, and on a random-access, diversity-enhancing basis.

It is obvious that one important 'institution' in the Transformational Society is the organisation and diffusion of communication/transportation/production technologies. I assume that all individuals are able to possess and utilise these servo-mechanisms and their successors. These items will function as interpersonal communication devices; monitors and diffusers of information about the state of the individual and the state of the system; and deliverers of individually desired goods and services, including food and health care.

If the underlying technology is actually 'cybernetic', then it is, by virtue of the very meaning of that word, 'self-controlling'. That is, I assume that individuals will not have to spend great amounts of time and effort in labour, physical or mental. I of course intend that individuals – separately or collectively, as appropriate – *can* intervene in the control process if they don't like what the cybernated technology is doing. The point is that they do not *have* to intervene, which is the case now, and has been throughout human history.

Hence, I assume that many political decisions – here, in the sense of 'collective' decisions – will either not be made at all (that is, a collective decision will not be made, the act being left up to individual preferences) or else they will be made 'cybernetically' (that is, according to the human-and-cyborganically-determined programme of the system). Where collective decisions are made – and they are to be made at different points in the system in accordance with the right of any individual to join with any other for any non-prohibited activity – the information technology of the time should enable many people individually to participate in collective decisions more effectively and rapidly than our present information technologies permit. This also should be the case, for those (hypothetically) few decisions that must be made globally.

Let me make a comment on this point. 'Cybernetic' political systems frequently give off the aroma of a political homeostasis. Many people, who seem to prize order over freedom, believe that it is possible and desirable to devise a stable social system, and they think that automated technology might help them do the trick.

I believe that intelligent cybernetic technology may be able to be used in the future to prevent many of the types of violence humans wreak upon one another at present, as I have indicated at several points before. I also believe that such technology can be used to smooth out those conflicts that result from genuine misunderstanding, lack of information, lack of effective participation in important decisions, and the like.

But I do not expect or want cybernetic technology – or anything else – to prevent those residual, though many, conflicts that result from individual genetic/environmental differences. That is the whole point of the Transformational Society: to provide a sufficiently 'secure' environment so that people can conflict and differ genuinely and freely, without 'master-slave', 'husband-wife', 'parent-child', ruler-ruled' and other inequitable and hierarchical distributions of power and status being morally, ideologically, religiously, legally or in any other way 'objectively' justified.

No, in a sense, I want the Transformational Society to be *thoroughly* 'political' – here meaning, conflictual rather than consensual. I do not want to establish any 'rules of behaviour' where they can be avoided, and I want the individual to have sufficient sources of *power* – which, after all, is the point of all this talk about technologies and institutions – *vis-à-vis* any other individual or group of individuals so that he can lead his own life freely, without being intimidated into conformity by threats or experiences of loss of security.

Freed from 'work', and hierarchy, individuals in such a future might very well spend almost all of their time in politics – making decisions, allocating resources, resolving conflicts and creating others. A world where all is politics. O Happy Day!

What should we do now? If such, in the haziest possible outline, is to be the shape of the Transformational Society, what should we do now to hasten it? Following are some of my suggestions. Since, in a sense, I think we should do all of them at once, they are in no order of priority.

Essentially, what we should do is identify those facts and trends in the present which seem to be providing the environmental reinforcements towards individual freedom and interpersonal tolerance, attempt to thwart those facts and trends which counter it, and create new perceptions or situations to further it.

1. Thus, we should do everything possible to create or interpret situations so that individuals will *experience* increasing self-fulfilment.

   (a) Protect those technologies of the present – such as the automobile, the bicycle and the motorcycle – which serve as enhancers of individualism, and resist retrogressive movements to de-individualising technologies – such as mass transit. This does *not* mean defending the present automotive system *in toto*; many features are destructive of human and environmental values and should be removed.

   (b) See that potentially individualising technologies of the present – such as television and the computer – are not used for purposes contrary to individual enhancement. This means countering many of the uses to which both of these technologies are currently put. Yet, the potential of both is towards individual freedom, in my opinion. One way in which we could further that aspect is to encourage more people to become 'literate' as active users of TV, the computer, etc, instead of mere consumers, as is now the case. Similarly, we might try to hasten the diffusion of coaxial cable television, being certain that its individualising aspects are stressed.

(c) Care should be taken to see that newly-developing technologies emphasise individualising features, where possible.

(d) The formal and informal educational systems of the present – including not only the schools, but also the mass media – should be encouraged to portray individualising experiences as enhancing, rather than threatening; to illustrate that the person who is non-violently different from the group is at least as valuable as the person who conforms; to reward the 'freaks' as well as the 'straights' for being 'true to their selves' and for trying out new selves.

The role of competition and violence in education should be reduced; and both the expectation and experience of diverse human attitudes and behaviour should be increased.

Educational curricula should be developed on a 'random access' basis, with different students learning different things at different times and in different sequences.

If there is one element that the school system should stress now, it is experience in making real choices. Rather than being taught 'facts' humans should be encouraged to identify and evaluate their own facts, and to formulate and choose from competing positions.

2. We should encourage 'developing' countries and parts of countries to bypass exploitative industrial technologies which seem to 'require' repetitive and boring labour, discipline, hierarchy, subordination, and obedience, and encourage them to develop cybernetic technologies in ways which are less negatively disruptive of existing human relations and environmental balances.

In the meantime, policies which will result in a more equitable distribution of available goods and services within and between current societies should be encouraged; those which seem to lead towards strengthened exploitative relations towards global consciousness, their irresponsible imperialistic characteristics must be resisted. In short, those trends which indicate a movement towards global consciousness coupled with local diversity and individualism should be encouraged; those which seem to lead towards strengthened exploitative relations must be discouraged.

3. Nationally, military expenditures must be very drastically reduced with all deliberate speed, if they cannot be totally eliminated, and funds diverted to:

(a) the development of person-centred fully cybernated technological and social systems;

(b) the development and human use of artificial intelligence;

(c) bio-medical research which will permit genetic engineering, the elimination of disease, the reversal, slowing, or end of ageing, and the like. Because of the great promise which it affords to further discoveries off this planet, a modest space programme should be continued. In other areas of central government spending, 'revenue sharing' and decentralisation have merit – even if not unambiguously so.

Decentralisation of decisions – ultimately, of course, down to the individual making his own decisions free from any group coercion – is to be encouraged in principle over centralisation, but where decentralisation results in, or is a disguise for, continued abuse of individual rights, then centralised decision making may be temporarily necessary, though deplored in principle.

Indeed, strong centralisation may also be appropriate in order to achieve equitable distribution of currently scarce material goods in developing countries or portions of developed countries. I am not certain that centralisation is the best possible prelude to decentralisation, however. While I expect that cybernated technology is likely to result eventually in greater individual freedom due to 'the logic of the technology', if cybernetic technology is developed with and for a strong central state — no matter how humanitarian or paternalistic it is — is it any more reasonable to assume that in the future the state will 'wither away' by getting stronger than it did in the past?

A personal note on 'development', 'democracy', and individual self-determination

In traditional societies conscious 'decision making' is relatively rare. There are few situations which require conscious choice. Everyone 'knows' what to do. In some of the relatively infrequent situations of novel choice, the individual can be expected to choose for himself by applying the obvious 'rules of the tribe', derived from the 'normal' situations, and probably come up with a decision which will satisfy most of the community (or, alternatively, dissatisfy most of the community — the point is that the community will, as a rule, be of a single mind about individual decisions).

Where conscious community decisions are needed, either an accepted leader 'knows' what to do as a consequence of his general participation in the characteristics of the community and/or his community-accepted and proven ability to act for the community; or in other situations (and more or less commonly in other communities) the community as a whole and/or its elders meets, discusses, and arrives at a mutual agreement concerning the novel problem.

Now if the above essentially typifies 'traditional' community decision making; if it characterises such decision making for most of humanity's most recent life (say the last 50 000 years); if it still characterises the experiences of most humans alive today; and if it lies in the very recent background of almost all humans' personal experience today, then it is no wonder that most people assume that consensus is 'natural', and that a 'mutual exchange of ideas' will result in both harmonious decision making and acceptable decisions, or else they assume that agreement is unlikely among heterogeneous persons and that avoidance of and hostility towards strangers, and a return to communal patterns, is the only alternative.

However, with the rise of new technologies, new experiences, new information, and the widening of breeding groups (situations that *may* have occurred generally first in the West) new rules for, and experiences of, human group decision making came into existence.

That is, an increasing number of common problems came to be experienced by persons who were not members of the same community and yet who had to reach mutually acceptable decisions. Out of this necessity, the institutions and styles of 'liberal democracy' grew: representation; political parties; separation of power; judicial and bureaucratic 'objectivity' ('a government of laws and not of men');

procedural and substantive rights; written, formalised 'constitutions'; and a decision-making technique called 'majority rule', arrived at by following precise, predetermined and written rules.

These, and other things, replaced 'consensus' as the technique for decision making in many areas of life for most 'developed' countries. But in many other areas within the developed countries – most notably the family, but also (in varying degrees) in fraternal, religious, occupational (especially labour unions), and similar groups – traditional consensual techniques have remained a frequent, if not exclusive mode even in the developed countries.

Modernisers and developers frequently seek to impose Western, majority-rule decision-making techniques on developing areas with only fitful success, and, as we all well know, in Europe and the United States too, those techniques are themselves increasingly being brought into severe question.

I have tried to show elsewhere that the continuation of the forces which brought liberal democracy (and other institutions of developed industrial societies) into existence are now rendering these institutions unnecessary if not actually obsolete, and it is this contention that leads me to recommend that we encourage developing areas also to avoid the Western forms and styles of political development as well as of economic development.

To summarise the argument, the decision-making modes of 'development' do not seem to fit easily into traditional patterns; they are not necessary in the Transformational Society; and they have no absolute value on their own that I know of that merits their forced transplantation.

From my point of view, the only 'development' worth striving for is individual human development – individual freedom. To the extent that 'economic development' or 'political development' leads to individual freedom – more rapidly and painlessly than any alternative – and it may – to that extent it is good. To the extent that it thwarts, needlessly postpones, or creates obstacles, then it is bad. I very much agree with Arthur Waskow's consistent insistence on living in the future *now* instead of constantly preparing, planning and sacrificing for it.

It is for this reason, too, that I strive not only to lead as liberated a life as I can within the institutions and expectations of the present, but also I try to test and redefine what I – and all others with whom I interact – feel the expectations of those institutions and roles are. (I do not try to foist my definition of the situations upon others, but rather to determine and act upon our mutual expectations. I did not say it was easy or that I was correct!)

Most importantly, I feel, I try to refuse to permit dependencies to grow up around me.

Since the future I seek is a world of maximum individual freedom, free from both exploitation of, and responsibility for, others, I try now to live as non-exploitative and – may I say – irresponsible a life as I am able (and, as I indicated, I try, gently, and lovingly, and tentatively, and inquiringly to push against what I and others feel are the limits of our abilities).

As father . . . as husband . . . as lover . . . as teacher . . . as friend . . . as son . . . as researcher . . . as preparer-of-a-paper-for-the-Rome-Conference-on-the-future . . . as 'futurist' . . . as – whatever else, I try to demonstrate that all I know and feel is what *I* know and feel — and it is always changing, always learning and unlearning, always growing and dying. I do *not* know *for* others, I do not know what others should or must know, and I do not know what others do know and feel — though I wish to try to learn if they wish to share. I have no light to shine before men; no secrets to unlock; no paths down which to lead others; no conversions to seek; no evil institutions to destroy; no motherfuckers to put up against the wall; no discipline to instill.

I want a future peopled by free, different, symbiotic, self-actualising individuals, and I think such a world is possible. But the way to it cannot be shown or prepared: it must be experienced (perhaps haltingly and sometimes painfully but often exhilaratingly) now, step-by-step, to nowhere and no time. It is neither there nor then. It is beginning now, and I do not see how the search for individual freedom and lovingly-interactive self-fulfilment can end. We no longer grow *up*, we grow. There are no limits to this growth. There are only continual transformations — like the wriggling snake shedding its skin — for (perhaps) ever.

For this reason, finally, I have purposely and consistently destroyed all attempts to establish a Movement around my (my!) ideas or person. I am no guru and refuse to be one. I do not want people to study my words or actions in order to discover the Truth in them. If there is any 'truth' in them — and I do not believe there is in any universal, objective, or eternal sense — then it is at best only true for me. Others must find the truth-for-themselves for themselves and through interaction with other enquiring selves.

For the reason that I do not seek to found a movement, neither will I join one — though I am quite willing to seek to share, in word and deed, in a great many other people's visions. But I personally feel that any search for, or insistence on, solidarity, community, orthodoxy, or dogma tends to impede — rather than facilitate — the Transformational Society of lovingly-interacting persons which I prefer and think possible.

The only way to be free in the future is to start being free now — non-exploitative and irresponsible; loving but self-actualising.

Whatever that means.

To you.

And to me.

Conclusion

I don't know what the world *will* be like in 30 to 50 years or beyond. I have tried to make clear in this paper how I would *like* it to be. If I have not been very specific in terms of structures, institutions, and present future-oriented policies, it was because I wanted to be as careful as I could in stating my values and goals, and because I do not want to prefigure the future with rapidly-obsolescing structures so much as infuse it with strategies and goals.

We have seen before that there are several other 'futures' which seem plausible alternative descriptions of the future I prefer, however.

One is the world of *ecodisaster* which many people see likely as a consequence of our irresponsible exploitation of finite natural resources, coupled with environmental pollution and people pollution. If such a nightmare becomes a reality, then the Transformational Society is not likely to result. Indeed, we can look forward to a world of perhaps unparalleled brutality and selfishness.

In contrast to this world — partly put forward as a way to prevent ecodisaster, and partly because many people feel it is a good way to live in and of itself — is the vision of the future that I call *ecoparadise*. This is the world which seeks a return to agrarian technologies and agrarian ways of life. Some people prefer, as we have seen, that we even go back to a pre-agricultural subsistence level of living.

There is very little likelihood that the goals of a Transformational World could be achieved under the communitarian — though certainly comfortable and secure — values of a neo-agrarian life, or the competitively rugged individualism of a society of hunters and trappers and gatherers of berries.

Some people fear that we may be able to avoid global ecodisaster only by advancing to higher forms of exploitative technology and institutions. They fear — as I do also — the global technocratic state, which is based upon the newest technologies and the most brutal of suppression. Perhaps the modern technocratic state will develop a modern equivalent to 'bread and circuses' and hence produce a 'friendly fascism' (in the words of Bertram Gross), but it will be a thoroughly programmed, thoroughly de-individualised world.

Perhaps, then, I am misguided when I want to try to develop a future world of freedom based upon cybernetic technology. It may just be too dangerous to try. The margin between total freedom and total slavery may be too narrow for human comfort. I just want to see us try to do things differently. I value trying to use new technologies and trying to create new institutions and values so that we can attempt to be humanly free as we have never been free before. I feel that it is more courageous — may I say, more human? — to try to do something new than to retreat into past forms of comfortable captivity.

Appendix: a note on the causes of crime*

If we insist on learning from the lessons of the history, then I believe we are bound to conclude that where there is no consensus there is no community, and where there is no community, there is no basis for law and the non-violent control of crime.

It may be that part of our problem is that we need to reconsider what we mean by 'crime' or 'deviant behaviour' generally. I do not believe that crime can be understood by reference to any particular behaviour *per se*. Crime is simply whatever a society defines or punishes as 'crime'. Crime is situationally defined, and in a rapidly-changing situation, we may be hampered by obsolete definitions as much as by obsolete institutions and technologies. Thus is any society, 'crime' can be eliminated in a number of ways.

One way is simply by redefinition. Many acts that have been defined as criminal can be redefined, and beyond that, we might even learn not only to tolerate but actually to enjoy witnessing and/or enacting these previously prohibited acts.

That is, instead of having society prohibit certain behaviour, we could instead ask that society help us understand what might be the genuine and significant, personally and socially negative consequences, if any, of a specific behaviour. This information would probably have to be given to us statistically rather than uniquely, and therefore we would have to determine for ourselves whether a statistical estimation, based on past aggregate data of situations analogous but not identical with our own, applies to us personally or not. It is possible to imagine designing a society where such information about the probable effects of specific human behaviour would be the only 'deterrent to criminal behaviour' which society would undertake at all, or in many instances.

Yet, we could also ask that an agency of society do more than that by giving us information which would enable us to become more tolerant of activities we previously found offensive. That is, rather than have the society seek to suppress acts which we feel outrageous, we could ask that society help us to suppress our own sense of outrage at these acts.

We could also try to design society so that previously criminal acts lost their undesirable feature. Thus, persons could continue behaving in the same way as before, but the effect would no longer be negative. For example, rather than outlawing or otherwise trying to prevent drunken driving, transportation systems could be designed so that the undesirable consequences of drunken driving — serious accidents, I suppose, though there might be others — are impossible or tolerably rare. For another example, instead of trying to outlaw theft — a concept meaningful mainly in a society based on private and inequitable ownership of scarce objects which are in high demand — technological and organisational methods could be invoked to design a production and distribution system by which everyone can have everything he wants freely and easily. For a final example, if biomedical sciences continue to develop apace, it may be possible to permit even murder if 'dead' bodies can be somehow rejuvenated, cloned duplicates were available, or the like.

The principle is clear, however: rather than try to prohibit certain behaviour, we could either learn (1) to tolerate it, (2) enjoy it, or (3) redirect it so that the behaviour can continue as before, but the effects are no longer evaluated negatively.

If, however, we remain determined to eliminate certain behaviour, I believe that we have two ranges of techniques available to us currently, or in the near future.

We can eliminate the social reinforcements that underlie the 'criminal acts'. In my opinion, 'criminal behaviour' (like all behaviour!) is not autonomously caused by (or the responsibility of) the individual 'criminal', as our laws and 'common sense' now assume, but rather is *partly* the result of the fact that certain individuals are caused by social reinforcements to act in ways that *others* define as criminal. To the extent that is the case (and I do not think that it explains all criminal behaviour, as you will see), if we change the reinforcement schedules appropriately, then we should be able to end the crime. This, then, might become the major focus of our attempts to create 'world order models': to design

* This is based on a portion of my 'Epilogue to *A Dowager in a Hurricane: Law and Legal Systems for the Futures,'* a multi-media presentation before the Citizens' Conference on the Administration of Justice, May 26, 1972.

social institutions which will eliminate the socially rewarding 'causes of crime' and will reward people only for behaving properly. [B.F. Skinner's *Beyond Freedom and Dignity* is an excellent attempt to do just such a thing on the national level. Presumably the same principles could be applied to international society as well.]

But we have a second (and not necessarily mutually exclusive) range of techniques available to us if we wish to eliminate 'criminal behaviour' because it seems that an additional contributing (some would say 'sufficient') cause of an individual human's behaviour (criminal or otherwise) is (1) his share in the genetic inheritance of the human race generally as received from his parents specifically, and (2) his own unique and genetically/environmentally determined biological structure and hence predisposition to act. From this perspective, then, a person's criminal act is genetically (or, more broadly, biologically) determined in interaction with his past and present environmental experiences. Hence, in principle and perhaps eventually in fact, we can direct a person's behaviour by altering — or predetermining — his biological make-up, chemically, electronically or surgically. There are many indications that we should be able to do this even 'better' (ie, more specifically and certainly) in the future.

So, to summarise: if we wish to end crime where there is no community, then (1) 'criminal acts' can be defined as non-criminal and either tolerated or enjoyed; or (2) the social environment can be restructured so as either to (a) lessen the impact or (b) remove the rewards of 'criminal behaviour'; or (3) the behaviour of 'criminals' can be 'corrected' by neuro-genetic surgery, or drugs, or electronic implantation and guidance, or the like. Only if none of these is possible (or desirable?) should a person or group be ostracised, deprived of their liberty or security, or murdered.

References

1. W. W. Rostow, *The Stages of Economic Growth – a non-Communist manifesto*, Cambridge University Press, 1960, 2nd edition 1970.
2. Max F. Millikan and Donald L.M. Blackmer, eds, *The Emerging Nations – their growth and United States Policy*, 1961.
3. Robert L. Heilbroner, *The Great Ascent – the struggle for economic development in our time*, 1963.
4. Irving Louis Horowitz, *Three Worlds of Development*, 1966, 1972.
5. Paul Samuelson, *Economics*, 7th edition, 1967.
6. J.A. Dator, 'The Protestant Ethic in Japan', *Journal of Developing Areas*, October 1966, p. 23.
7. Hajuine Nakamura, *The Spirit of Capitalism in Japanese Buddhism.*
8(a) S. J. Bodenheimer, 'The ideology of developmentalism: American political sciences paradigm-surrogate for Latin American studies', *Berkeley Journal of Sociology*, 1970, pp.95-137;
8(b) S. J. Bodenheimer, *Politics and Society*, May 1971, pp.327-357.
9. Ezra J. Mishan's *The Costs of Economic Growth* was not the first, but was one of the more strident and influential voices of dissent. Since the publication of his book in 1966, there has been a swelling flood of similar protests – so many, in fact, that Mishan's complaint is rendered almost comically out-of-date by its tone of self-righteous indignation. His perhaps once-novel and unorthodox observations about the true 'costs' of economic development have been so loudly repeated – indeed, parroted – recently as to have themselves become shibboleths on a par with those which he intended to put to rout.

In a recent lead review in the 'Book Issue' of *Science* (May 11, 1973, pp.580-582), John Platt opened his evaluation of five recent and related books with the observation that 'Something like a world survival movement may be what we need next, and in fact it may be larger already than most of us realise' (*Ibid,* page 580). The books which Platt reviewed are indeed of a piece – though not without their significant differences. (They were Falk, *This Endangered Planet*, Wagar, *The City of Man: prophecies of a world civilization in twentieth century thought*; Goldsmith *et al, Blueprint for Survival*; Brown, *World Without Borders,* and Gabor, *The Mature Society* – Gabor's book is almost out of place on this list, however). Platt listed other books that had appeared previously which he thought fit into the same general mould: *Man's Impact on the Global Environment; The Limits to Growth; The Radical Alternative; Without Marx or Jesus*; and Sakharov, *Progress, Coexistence and Intellectual Freedom.*

He might well have mentioned others: James Ridgeway, *The Politics of Ecology*; Paul Ehrlich, *The Population Bomb*; Paul Ehrlich and Richard Harriman, *How to be a Survivor*; Charles Reich, *The Greening of America*; Philip Slater, *The Pursuit of Loneliness*; Theodore Roszak, *The Making of the Counter Culture*, and *Where the Wasteland Ends*; Murray Bookchin, *Post-Scarcity Anarchism*; Garrett Hardin, *Exploring New Ethics for Survival*; Paul Barkley and David Seckler, *Economic Growth and Environmental Decay*; and well, where should I end? There have been so many!

All of these more or less share the commonalities identified by Platt: 'that our problems are (1) multiple and interacting, (ii) global, and (iii) urgent, on a time scale measured in years rather than decades' (*loc cit*). Platt also says that they identify the problems similarly: 'peace keeping, the rich-poor gap, the exhaustion of non-renewable resources, population growth, and pollution and damage to the ecosystem' (*loc cit*) because both 'our old establishment programme' (the American Development model?) and 'our old revolutionary arguments' (such as the Bodenheimer critique) 'are rapidly [being] overtaken by the spreading planetary networks of television, tourism, technology, and trade, nothing less than 'an integrated global analysis' and integrated global solutions will suffice. The volumes he reviews, plus most of the ones I also listed (and many I did not) all attempt to launch a systemic, global attack on the cluster of problems he listed.

10. Rachel Carson, *Silent Spring*, Boston, Houghton Mifflin, 1962.
11. Paul Ehrlich, *The Population Bomb*, 1968.
12. Margaret and Harold Sprout, *Ecology and Politics in America: some issues and alternatives*, General Learning Press, 1971.
13. Yi-Fu Tuan, 'Our treatment of the environment in ideal and actuality,' *American Scientist*, May-June 1970.
14(a) Tony Wagner, 'The ecology of revolution' in *Ecostatics*.
14(b) James Ridgeway, *The Politics of Ecology*, New York, E.P. Dutton, 1970.
15. Donella H. Meadows, Dennis L. Meadows, Jørgen Randers and William W. Behrens III, *The Limits to Growth*, Universe Books, New York, 1972 (a report for the Club of Rome). A detailed critique of this report by the Science Policy Research Unit, University of Sussex, UK has been published in *Futures*, Vol 5, No 1 (February) and No 2 (April), 1973.
16. Richard A. Falk, *This Endangered Planet – prospects and proposals for human survival*, New York Vintage Books, 1972.
17. Gareth DeBell, *The Voter's Guide to Environmental Politics*; John Mitchell and Constance Stallings, *Ecotactics: The Sierra Club Handbook for Environment Activists*; Sam Love, *Earth Tool Kit*, New York, Pocket Books, 1971.
18. Gus Hall, *Ecology: can we survive under capitalism*, New York International Publishers, 1973.
19. R. Giuseppi Slater *et al*, *The Earth Belongs to the People – ecology and power*.
20. Charles A. Reich, *The Greening of America*, 1970.
21. Phillip Slater, *The Pursuit of Loneliness*, 1970.
22. Theodore Roszak, *The Making of the Counter Culture*, 1970.
23. Denis Goulet, *The Cruel Choice – a new concept in the theory of development*, 1971.
24. Erich Jantsch in a collection of radio transcripts which comprise G.R. Urban, ed, *Can we survive our Future?*, 1971.
25. John Lukacs, *The Passing of the Modern Age*, 1970.
26. Herman Kahn, *Things to Come*, 1973.
27. Kenneth Boulding, *American Behavioural Scientist*, January-February 1972.
28. Andrew Shonfield on 'Change and Social Good' in ref.24.
29. *Science News*, 14 April 1973.
30. Joseph Gusfield, 'Economic development as a modern utopia' in David Plath, ed, *Aware of Utopia*, 1971.
31. Northrop Frye in Frank Manuel, ed, *Utopias and Utopian Thought*, 1967.
32. John McHale, *The Future of the Future*, New York, George Braziller, 1969.
33. René Dubos, 'Humanising the earth', *Science*, 23 February, 1973, p.769f.
34. William Glasser, *The Identity Society*, 1972.
35. Edward Heineman, *Freedom and Order*.
36. John Platt, 'Social traps,' *American Psychologist*, August 1973.
37. Arthur Clarke, 'The mind of the machine', *Playboy*, December 1968, p.122f.

Reflections on the relationship between the individual and society

William Simon

Let me begin with a question, whose answer remains a constant and constantly unexamined assumption in most futurist thinking: Why must humankind survive? The question is really two questions. The first is the absolute version of that question and touches on the concept of the future as such. What is it that constrains us to commit so much of our lives and resources in order to maintain an "enterprise" beyond the span of our own lives, those of our children, and even those of our grandchildren? For most of human history and possibly for most of humanity alive at this moment the question in this absolute form was never asked; life was too much tied to the implicit order and periodic disorder of nature. This, incidentally, was the ideal environment for the development of *symbiotic* relationships, which have been defined as the *unintended* cooperation of *unlike* species. But the very history of humanity, of what has been called traditional societies (or pre-industrial societies, though these are clearly not synonyms), has been the movement away from a predominance of symbiotic relationships compared to those characterised by sociocultural regulation, by *rationalised* (though not necessarily rational) modes of cooperation.[1]

Concepts of man and scenarios of the future

With a rupturing of the continuity of traditional society, the "why" of humanity's survival was dramatically placed on the social agenda. With the messianic tradition of Judaic thought and the chiliasm of Christian teleology a partial answer was possible. This, of course, was the very meaning of Weber's observation that the Protestant Reformation, in effect, "destroyed the walls of the monastery and transformed the world into a monastery". We were to prudently seek prosperity and with equal prudence use the fruits of our prosperity in accordance with the dictates of divine intent. The secular response, more appropriate to the scientific-technological ethos of industrial society, was tied to notions of progress, service to history, the furtherance of sociocultural evolutions and, implicit in all of these, the perfectability of humankind. Both of these responses, the religious and the secular, became not only the basis upon which both individual and collective goals and aspirations could be fashioned, but also became the basis upon which critical assumptions (and, implicitly, limitations) about what persons and society could become could be supported.

Characteristically, the secular notion of sociocultural progress in the Western world was largely shaped by the problematic nature of survival potentials, by the near constancy of deprivation in terms of the most basic human needs, such as nutrition, shelter, etc. Much of the conventional wisdom of the Western experience was predicated on the assumption that deprivation or near deprivation was the normal state of affairs and prosperity the unexpected and pathogenic moment. (This, for example, was very much the position of Durkheim who tended to view prosperity as the point of greatest societal vulnerability to anomie.) Prudence, hard work, and achievement, resting upon a commitment to the future in terms of salvation for the religious and historical validation for the secular, tied the very conception of humanity to an ever shifting future point in time; much as "personkind" in pre-industrial societies was bound to the past, bound to a sense of fidelity to tradition. Both of these abstraction-dominated perspectives ensured that activity in the present, however unrewarding and self-denying in itself, served the collective enterprise. The dominant perspectives organising a sense of self were developed around a limited number of social metaphors. Even family life was organised in this way, which is the very meaning of the invention of childhood and the subsequent invention of adolescence and youth: the child-centred family was, and is in actuality, in essence the societally-oriented, future-oriented family.

The emergent forms of post-industrial social life have called many of these conceptions into question. The experience of relative affluence over the length of a generation for major sections of most industrial societies has provided a critical testing of most prior conventional wisdom — wisdom, again, predicated upon preparing us to live lives of adversity and deprivation. The very beginnings of this era were defined by John McHale as the point at which societies began producing human beings who could be expected to endure longer than the things they produced. Previously (and currently for most of humanity living on the planet), things stood as more enduring markers on the social horizon and as such frequently became both the meaning and measure of human existence. Some part of this is implicit in Marx's notion of "the fetishism of commodities". But as industrial and post-industrial technologies alter this relationship by making the world of things less enduring (how many of us will continue to live in the same communities, even if we do not physically move?) and less expensive. All at once, it becomes harder among the affluent strata for the older to justify and the younger to organise their lives in terms of the objects they accumulate. The aspirations that once served as pole stars for men increasingly fail as concrete, existential realities; existential realities that, as Durkheim anticipated, lose their ability to organise behaviour as they become pedestrian and banal. Despite the strident, Jeremiah-like warnings of the "limits of growth" ideologists (in their multiple guises), larger and larger segments of the population of such industrial and near post-industrial societies experience the discomforts of having to do what the whole of their prior training ill-equips them to do: live with a strong sense of the present.

The children of the affluent, being, as it were, born on the far side of achievement and success, tend increasingly to develop an immunity to the oppressive sense of the future that tyrannised so many of their parents. (Even their parents increasingly require the invention and reinvention of alternative therapeutic modes to be temporarily reassured of their own reality and meaningfulness. Indeed, futurism, which rapidly takes on many of the forms of a social movement, may function for many as just that kind of therapeutic mode.) While what might be called the anomie of affluence engenders a devaluation of the future as a constraining force in personal and social life, those still victim to the anomie of deprivation in both industrial and pre-industrial societies, being increasingly exposed to the imagery of post-industrial life, also learn to organise their demands in terms of an equally strong sense of the present. The rhetoric of the moral value of sacrifice and denial in order to preserve the viability of human society beyond the farthest vision of contemporary humanity should meet with less and less enthusiasm.

Let me offer something of a parable in summing up this new potential rejection of the future. I have among my acquaintances a young poet who was also what we in the USA have come to call a "speed freak" (ie, one strongly committed to heavy amphetamine use, though he also freely used other drugs as well). Once, in conversation, I reacted with what I hope is an uncharacteristic conservatism. I suggested that he was going to destroy himself prematurely. His reply was that I could not understand him largely because I still operated with an outmoded sense of time. As far as he was concerned, both his body and the world around him could only exist meaningfully for him as sources to add to his mind's capacity to experience things as widely and deeply as possible. He added that his life span might ultimately represent only a small fraction of mine or even his own as an abstract potential, but if during that short span he experienced more things and more things more passionately than I with my more prudent approach, then it would have all been worthwhile. The question he was raising was the question I began with: Why must humankind survive? Frankly, then as now, I had no answer. At least no answer that did not rely on shop-worn images of the divine or moral platitudes that, having no viable foundation, also had little necessity.

What I am trying to suggest, partly as McHale has done already in his *The Future of the Future*, is that the future is not a given, but a social invention that cannot be taken for granted.[2] This further suggests two very specific concerns. The first of these is that we must learn to build into our models or scenarios of the future a concern for the future as a variable that will vary with the conditions of social life. In this context, our very preoccupation with the future may have more to do with the politics of the present than with the science of the future. The second points to one of the basic cleavage lines around which the political struggles that will shape the future will form: different time perspectives. Just as perhaps during almost every moment in human history, "the past, the present, and the future compete for breathing space within the same social moment", so the political struggles of different interest groups and ideological perspectives, both within and between societies, will reflect different commitments to a sense of time and our obligations to it.

Among the most difficult tasks for an individual is to conceive of a universe without people, or at least without people as the central actors. It is almost as difficult to conceive a universe with a version of the human radically unlike our own. Which brings us to the second or relative form of our initial question: Why must humanity survive in some form that we might recognise as comparable to our own experience? As a predominance of symbiotic relationships gives way to a predominance of rationalised relationships, as the "disenchantment of the world" that Weber described follows the complex division of labour (Durkheim's "organic solidarity") and bureaucratic modes of organisation required by contemporary technologies, cultural institutions face a crisis of their own impermanence. The more manifest impermanence of things begins to call into question the permanence of humanity. Recognition of the essentially protean character of social orders, the recognition that one's time and place were not necessarily the embodiment of some divine intent nor even necessarily a crucial step on the road to some ill-defined point of historical culmination has created problems of commitment both to the social order and the future.

Critical to most conceptions of the future, as well as to the activities performed in its name, is the sense that the occupants of some future time could comprehend the meaning and motivation that informed our activities, accomplishments, and monuments. History, the ultimate act of reification, would, as it were, stand as our judge. Despite the fears of many, such as Weber, that the inherent pragmatism (if not scientism) of industrial bureaucracies had set loose a "disenchantment of the world", personkind's capacity (if not need) for enchantment and for mystery persists. It is just that the focus and raw materials, so to speak, of enchantment and mystery shift and are subject to reordering. The line of thinking that ran from Saint-Simon to Comte to Durkheim, that asserted that the social sciences would re-place traditional religion as major norm-generating and validating institutions may not have been in error. The social sciences, with relatively few exceptions, become the source of the new images and icons of enchantment and mystery, ie, the critical assumptions that are simultaneously beyond testing, and universal, as the focus of contemporary enchantment becomes humanity itself. From a universal divinity we shift the focus to a universal model of the human.

A universal — transcultural and transhistorical — model of the human with a core cluster of needs, abilities, and disabilities that articulate with and set limits for alternative forms of social life appears essential to the practice of social science — Marxist and non-Marxist alike — and to most futurist thinking. This conception of personkind's universal nature undoubtedly gives rise to attractive, if not desir-able, politics, eg, "the brotherhood of man", but also tends to give rise to a dis-guised ethnocentrism and consequent conservatism. We learn to live with a protean conception of social life by retreating to a seemingly fixed conception of the human.

The various universal models of humanity that we are offered vary. We may argue as to whether it is innately cooperative, competitive, or both, but in their most general form the models tend to be celebrations of an image of the human with which we tend to feel most comfortable. Elsewhere, I have described the tendency to utilise the past, to use history, to reassure ourselves of our permanent place in a permanent species. The latent desire to "colonise the future", for which much futurist thinking can be validly criticised, has for many as its most fundamen-tal aspect *not* the desire to organise ideologically our response to the future in terms of perpetuating different existing institutions or institutional forms, but to perpetuate a limited conception of the human itself, a conception (or conceptions) shaped by the sociocultural moment. We may have to begin to face the fact that as the social order changes so do people, and that these changes occur on the most fundamental levels.

As David Fischer observes, we tend to reorganise the historical record in order to make its actors (our ancestors) comprehensible to us in terms of our sense of the human and human motivations. In fact, our predecessors, including our fairly recent predecessors, may represent profoundly divergent versions of the human, divergences as great as we might expect with any extra-terrestrial forms of intelligent life to come in from outer space. There is too much variability regarding critical matters of life and death, not only in the historical record, but (for better or worse) within the contemporary human experience to sustain such a position.

Psychologists and psychiatrists continue to offer us universal models of the human, with common needs, common disorders, and that involves a universal language that usually translates down to a given generation's best hopes and fears about itself. Child development researchers, following Piaget, continue to seek for an understanding of the organism's implicit agenda for optimum development; rarely, however, pausing to ask: Optimum for who? Or for what? There are times when I am convinced that the predominant weight of Western social science points towards a utopian vision of a world where all institutional forms are committed to maintaining the growth of healthy children, which, in turn, will be demonstrated by their ability to raise continuing cohorts of healthy children. Even in our present state of secular confusion, we can find more interesting uses for life than that!

Lastly, the ethnologists document phylogenetic continuity with parable-like exercises in anthropomorphic thinking, ie, imputing human attributes and responses to non-human organisms. Thus, they are imputing an intended (ie, evolved) set of limits which is in fact unintended (ie, arbitrary) to what the individual and society can become, that returns us to the absolute and somewhat more abstract rule of the symbiotic needs of inarticulate nature, ie, the ecosystem.

We learn to accept the realities of protean society by clinging ever more passionately to a constant or near constant view of the human. The essence of this new, man-centred "enchantment" reflecting what I would term the fallacy of "sociomorphic" thinking (ie, the imputing to the individual organism that which properly is a function of social life), allows us to accept uncritically as constants, as sociocultural universals, or "natural laws" for social life that which may, in fact, represent little more than the outcomes of the sociohistorical moment. Let me be more specific by way of example: Men (and if not men, surely women), we are told, experience a need to engage in reproductive behaviour. As one who has done a fair amount of research on sexual matters, I would argue that there is no data to suggest that man in his organic constitution brings to social life an impulse to engage in reproductive behaviour or experiences his or her sexual activity as the expression of that need. The left touts cooperation and the right competitiveness as basic mechanisms. Territoriality and aggression on the negative side or self-actualisation (whatever that might really be) on the positive side are additional examples. So are the sex role stereotypes that condemn half the globe's population to second class status. In each case, to the degree that thinking about the future

145

deals with man, which except in the most abstract form is not very often, one almost invariably finds the invocation of one or more of these as unexamined givens. Again, I would assert that there is precious little systematic evidence to suggest these are borne by the individual organism in ways that significantly organise behaviour or add significantly to "the social construction of reality".

One of the clearest examples of sociomorphising can be seen in the use of the fashionable concept of "alienation", a form of social pathology to which urban-industrial societies (and particularly the intellectuals they contain) are alleged to be especially vulnerable. This position requires a sense of some ultimate or original nature that given social contexts repress, deny, or distort, much as the conception of "false consciousness" implies the existence of "true consciousness". For Marx, alienation had roots beyond the cruelties that attended the domination of "living labour" by "dead labour", but rather they rested in the conception of humans being ontologically-producing, cooperative organisms made to despise themselves, as well as adding to the condition of their own oppression by their participation in the one activity that celebrates man's distinctly human nature. For Marx that which may have only been a stage in human evolution becomes the significant meaning of that process. Again, there is little evidence for this beyond a selectively edited version of the human record, itself having little depth in terms of human lives lived, as well as being a way in which the dead oppress the living.

The danger of the sociomorphic fallacy is that it reverses, or at least confuses, our sense of causality, leading us to see the person not as the constantly changing end product of a continuing process, but as an organism whose very constitution organises and constrains that process. One can understand the need for this kind of thinking, as it provides for the illusion of continuity, particularly continuity of meaning, in a period of rapid social change, in a universe that affords few constants beyond those we create for the moment.

To really accept the idea of a protean society is also to accept the idea of protean people. Both are inseparable aspects of the same process and the same reality. In terms of thinking about the future, we must learn not only to ask how the future will affect us, but also how it will transform our children. Our models or scenarios of the future must incorporate the human as a variable, in the multiple sense in which that term may be used. Those committed to designing future social orders must also understand that to the degree that they succeed, they will also be creating a new version of humanity. If nothing else, the future, which is a problematic pole star for many of us, becomes the oppressive, natural order of things for its inhabitants.

My original assignment was to consider the areas of conflict between the individual and society. I have begun with this rather lengthy introduction in order to suggest the meaninglessness of the formulation of that assignment, as well as some of the reasons why we continue to readdress it, as if all social thought were little more than a footnote to Hobbes and Rousseau. This is not to say that conflicts within human societies will not continue, but rather that such conflicts can only be understood in terms of concrete social contexts and specific human experiences and not in terms of abstract conceptions of Individual and Society (each capitalised as if suggesting core permanent qualities for both). Let me turn next to a brief projection of one version of what social structure and human personality might look like in the post-industrial world whose imminent birth is currently being heralded; then to consider an individual-society conflict, but in terms of alternative versions of the human.

We began by asking: Why must humankind survive? To that we had no answer. The second question was: Why must humanity survive in ways that we would recognise as being close to our sense of the human? The answer to that is that it probably can not.

Temporary societies

The possibility of a social life that can be holistically experienced (or even fully comprehended) becomes impossible or nearly impossible in a social order that is increasingly organised on a global and potentially super-global basis, though many of our traditional values insist that this is precisely the arena within which serious individuals realise their destiny. The earlier image of mass society tended to be more frightening to intellectual aristocrats who had this commitment to serious work ("leaving your mark upon the world") than it did for more ordinary people. As the scope, size and density of social life expands, the need and ability to organise lives in more personal terms increases. What also would have happened is the potential release of persons – particularly serious persons – from the ultimate totalitarianism that has characterised most of human history: the need to serve society, to live at ease in a world where there may not be any purposes larger or more compelling than their own. A world begins to emerge where things, relationships, environments, and even other people become freely substitutable and the closest things individuals may be able to treat as a constant they may have to find within their own changing character: a temporary society containing temporary people – as if we have ever been anything else.

Important elements in shaping post-industrial society might be seen more compactly in Table 1, which we will briefly elaborate before going on to the implications of such changes for "individual society" conflicts. The table, among other things, allows us to see these trends in a historical perspective, which should serve not only to point out the dramatic nature of impending change, but also to suggest how equally dramatic have been the changes that have brought us to this juncture. The table, simplifying as all such devices must, considers three categories of history: pre-industrial, industrial, post-industrial. These are hardly our invention; their sources – particularly the substantive content under each – are numerous. One intellectual debt, however, that must be acknowledged is that owed to Professor David Riesman whose works have substantially influenced the author. Anyone who has read *The Lonely Crowd*, a work that while flawed by a number of defects – most of which were made apparent not by insight, but by time – remains an extremely important work of continuing viability, must be aware of the considerable overlap between Riesman's conceptualisation and the one presently offered.[3] Clearly, Riesman's "tradition-directed" man found himself most at home in pre-industrial societies, "inner-directed" man being appropriate to the early and middle periods of industrial societies, while "other-directed" man and the projected

TABLE 1. ELEMENTS IN SHAPING POST-INDUSTRIAL SOCIETY

		Pre-industrial society	Industrial society	Post-industrial society
1.	Dominant social organisation	Caste (vertical)	Class (vertical)	Life-style (horizontal)
2.	Definition of social role	Categoric	Career	Episode
3.	Commitment to social role	Permanent	Adaptive	Temporary
3	Role performance	Ritualised	Playing roles	Role playing
5.	Basis of conformity	Natural identity	Code of conduct	Personal justification
6.	Role variation	Sex role	Sex roles	Personal style
7.	Response to deviance	Expiatory	Restitutive	Therapeutic
8.	Internal control	Shame	Internalised conscience	Sense of well-being
9.	Character structure	Ego/ego ideal	Ego/super-ego	Ego-reference groups
10.	Development of self	Sequential (age-graded)	Developmental (differential)	Variable (optional)
11.	Temporal frame of reference	Past-oriented	Future-oriented	Present-oriented
12.	Individual orientation	Fidelity	Achievement	Approval and/or experience
13.	Imagery of self	Social metaphors	Social metaphors/ personal metaphors	Personal metaphors

"autonomous" man offer plausible outlines in later industrial society as well as post-industrial society.

Before elaborating upon this table, a brief cautionary note is necessary. Early in the present essay we commented upon the futility of trying to design too specific a scenario for the future. Almost everything that has gone before and in particular the table just presented, while representing what we feel are highly probable outcomes, do not necessarily lead to particularly concrete portraits of either a society or an individual or modal type appropriate to that society. These aspects of social or personal life can contain a wide variety of contents and can be organised in an equally wide variety of ways. It may be more effective to indicate what is not likely to "fit" with the emerging future than to indicate with any precision what will eventuate in those that do "fit".

Dominant social organisation

Pre-industrial societies were predisposed to fairly rigid, caste-like organisational systems, being highly ascriptive. To know something about an individual's initial social location (usually the family into which he was born) allowed one to predict almost everything we need to know about that individual.

Industrial society takes as its basic building block social class, with homogeneities based upon common class position. Being more linked to achieved rather than ascribed status attributes, it allows for greater mobility. Knowing something about an individual's social class background still allows for reasonable prediction, but one of far less certainty. Both, however, remain essentially vertical societies.

Post-industrial society has as its basic unit the more voluntary and perhaps vaguer concept of "life-style commitment". What you are is here a function of what you want to do, where and how you want to do it. Here, consequently, the ability to predict begins to break down. While some regularities may occur as expressions of biological limitations, this more horizontal social order affords limited predictability either across time or between sectors of life (eg, "poor jobs" need not be predictive of "poor" or permanent life-styles).

Social roles

Pre-industrial society tends to be *categoric* in the sense that beyond adulthood, however that is defined, little variety occurs and there is essential consistency in role performance in different institutional settings or moments. The concept appropriate to industrial society is *career*, suggesting a series of stages or episodes linked to each other, yet only probabilistically; a trajectory involving the achieving or maintaining of status against which one's activities can be assessed and, more importantly, where one can win and lose (advance or fall back). Post-industrial society resembles industrial society in terms of stages or episodes, but stages or episodes without a trajectory, without a compelling narrative theme, possibly with only the most abstract sense of order beyond opportunity, ability, and need and/or desire.

Commitment, performance and conformity

The definition of roles in pre-industrial society leans heavily towards the permanent ("it was, it is, it will always be"), giving rise to what appears to the observer as ritualised behaviour because there is little variation between different individuals in the same role or the same individual performing in that role over time. The actor in a pre-industrial setting conceives of his role un-selfconsciously because it is defined as a part of his very nature (eg, "I do not hunt, I am a hunter").

Surrendering after fierce resistance to the fact of social change, as well as the creation of a complex and larger society, the definitions of roles in industrial society admit to more flexibility, are seen as adaptive in the sense that the actor can substantially add to the role (playing it well or not so well), and that this quality of performance determines what subsequent roles he or she will play. Moreover, the

149

styles of given roles are learned, partly by one's learning a role-specific code of conduct that is sufficiently abstract to create uncertainty with regard to appropriate application. What is important, however, is that once learned they tend to be regarded by the actor as part of his essential nature.

Post-industrial society, once again, resembles patterns of industrial society but with a sense of continuity substantially reduced; lacking an obsessive commitment to traditional status distinctions (which provides the sense of progress or failure), roles may appear and reappear without order, which gives them their temporary character. Further, given temporary commitment, they tend not to be internalised as part of the actor's character, but rather something that is adopted much as a theatrical actor assumes a role, as something to experiment with and experience for a while. The selection of roles by the individual is largely a function of his or her needs or interests at the moment.

Variation in roles

Much of this may be clarified if we look at the case of sex roles. Sex role expectations in pre-industrial society tend to prescribe the expected behaviour styles across a range of social activities and to describe in narrow terms all but a few exempt members of a sex. The sex role expectations of industrial society also maintain a core of behavioural styles that are common to most members of a sex, but it also tolerates a large number of options (the options for one sex still being different from those of the other sex) as long as the options are related to and consistent with a particular role or status. For post-industrial society almost all "sex role attributes" except those that are immediately tied to biological differences (and these turn out to be a very small number) tend to cease being gender-specific, cease to be part of a continuous general character, and tend to be adopted and shed as roles and situations change.

Deviance or flawed role performance

For pre-industrial society, deviance (which being unnatural provides for virtually no plausible motives) requires an expiation that purges both the offending individual and the uniform society of that offence. Deviance in this context is a form of contamination requiring ritual purification. Given the uniformity with which acts and performers are evaluated in this setting, responses to flawed performances are both immediate and universal. As a result, shame — or fear of being ashamed — which depends upon external response, becomes the mechanism that prompts individual conformity. Indeed, shame is a risk run by more than the individual actor, but extends to all members of a group to which the deviant belongs.

For industrial society, the "flaws" involved in the commission of deviant acts are essentially functions of the individual who must select from a complex of norms, many of which may be inconsistent with one another and are far less likely to be universally applicable or even recognised. Marking his less direct, more fragmented (or specialised) attachment to the community, punishment in industrial society is conceived of almost contractually, as an act of repayment — "paying one's debt to society". Conformity is achieved through the development of an internalised conscience. In the absence of long established and well set criteria or a basis of universal applicability, the individual in industrial societies must be trained to judge himself or herself.

The increased segmentation and individualisation to be expected of the post-industrial individual makes deviance both harder to define and less threatening to the collective enterprise. And for just these reasons it is likely that it will be treated as something to be corrected, with the society intervening only when potential harm to the actor or the community becomes apparent. The response may well resemble that of the "psychosocial moratorium" that contemporary Western societies — however unevenly and inconsistently — extend to middle class youths. Similarly, centralised mechanisms of control can also be relaxed, with control resting either upon group approval (a small and specialised group) or on a sense of well-being ("some things are not done because they represent too much of a hassle").

Aspects of self

Finding his ultimate sense of legitimacy from a strong sense of the past or tradition, pre-industrial man has his sense of worth determined by the fidelity with which he approximates traditional patterns of behaviour. The very structure of surrounding social life facilitates this by patterning the transition or change points in his life so as to make them minimally responsive to personal variability. A male child at X years becomes Y, and that obtains for all males at X years — the smart and not too smart, those big for their age and those small for their age, etc.

In industrial society, however, where subsequent stages in one's career are only probabilistically related to earlier stages (being X as a child provides no absolute guarantee of being Y later), much of one's orientation, at any one moment, tends to be shaped by its future consequences. In this context, the individual's sense of his or her own worth depends on his or her sense of ultimate achievement, where even realised achievements have their meaning determined by their relationship to even more ultimate achievements. And while there is a general sense of a common sequence of life-cycle stages, there is considerable room for individual and group variation; whole stages may be skipped — particularly in the name of ultimate achievements.

Post-industrial man, we can expect, will be more rooted in a strong sense of the present reinforced by the lessening of the hierarchic evaluation of given social roles. The post-industrial individual's sense of his own worth will derive more from what he or she is rather than from what he or she might or must become. The essential value of what he or she is doing derives from the experiences intrinsic to the activity, rather than the instrumental value of that activity for the realisation of other ends — ends projected into an uncertain future. Individuals of the post-industrial era will want to be paid as they go. This does not mean that there will be a total lack of commitment to the future or an inability to engage in labours the conclusion of which must be projected into the future; rather and merely that the conditions of such commitment and such labour will also be judged in terms of their immediate impact on their contemporary existences.

The imagery of self

Lastly, the language that pre-industrial actors have at their disposal for describing themselves is almost entirely non-individualised; they cannot see themselves apart from the social world within which they live. For them there is only external reality and its requirements, a world with limited uncertainty because what is expected of them is also defined for them as part of their very natures. Persons in industrial society are acutely aware of themselves as individuals, but their realisation of their own identities depends on their realisation of social goals. Where there is both social and personal language available, there is also an assumption of a high degree of correlation between the two – social metaphors being more enduring than the personal ones. They can be described in the language of socialisation as those who "internalise the external". For post-industrial societies this order may reverse itself: a sense of commitment to personal language being stronger than social metaphors, they may be described as those who "externalise the internal"; being less concerned with satisfying the world, they ask whether it can satisfy them; being potentially less concerned with their ability to conform to the expectations of the surrounding community, they are free to ask: Does it meet my requirements? Their first loyalty is to themselves and that sense of self becomes linked to the social in indirect, complex, and highly variable ways. Once again, this does not automatically imply an inability to experience loyalty to others, to organisations, or even the society, but that such loyalties will be limited at the point of a denial of self (for example, a situation where parents can no longer indict their children with the burdens of their "sacrifices").

Individual and society: the future of politics

With the foregoing as, if nothing else, a sense of my best guess as to where the human enterprise appears to be heading on the most general level, I will now turn to the question at hand: conflict between the Individual and Society. Or, if you prefer, Society and the Individual. Implied in this frequently raised question, as I have already suggested, is some fairly abstract notion of identities that not only can be defined independently of one another, but that also are often defined in terms of contradictory needs or interests. In contemporary imagery, the best example is perhaps that of Freud, who clearly saw society or civilisation as a necessary imposition on the individual's phylogenetic heritage – an imposition the organism resists despite its basic dependence upon societal regulation and that is maintained only at the expense of a continuing sense of frustration and rage at one or another level of consciousness. The same model may be found in much of the practice of current behavioural science, including many behavioural scientists committed to playing the "futures game" (or is it "futures market"?), particularly those concerned with humanity's alleged capacity to create an "inhuman society".

An alternative, and equally deficient model eliminates as meaningful the individual-society conflict – a position that has appropriately been described as "an over-socialised conception of man" – where the individual actor is defined as being essentially what social life requires her or him to be. This can be seen in the functionalism of current Western social thought (though many of the functionalists manage to sneak in some elements of "immutable" humanity via the definition of social system needs) and the "up-beat" revisionists of Freud, such as Fromm or Erikson or May. The same essential position may be found in the social sciences of those countries where Marx's critical spirit is transformed into a daily guide to positive thinking and positive living so that grey figures in grey bureaucracies might lead us towards the sombre joys of socialism.

The latter position, despite its ideological utility for the maintenance of the status quo (any status quo — left, right, or centre),[4] fails for the following reasons. First the very size, density, and complexity of all but the most irrelevant forms of societal organisation make it virtually impossible to experience participation in anything resembling the larger collectivity. The nostalgia for a lost and usually mythic conception of community may persist, but few of us born to or encapsulated in the culture of urban-industrial societies (and increasingly for all other societies) will be allowed to go home again or even, as it were, be allowed to stay at home. The opportunity for a sense of shared experience, which is the foundation for the most fundamental kinds of consensus, will describe small and relatively impermanent social groupings as the units which both organise and provide the context for experience grow larger and larger. Clearly, for example, the idea of social class — whether it is defined as near identical relationships to the means of production as in Marx or the sharing of comparable life chances as in Weber — in no real contemporary sense forms the basis of mutual identities or even solidarities, as much of recent Western political experience amply demonstrates. Similarly, even when we share the same experiences at the same moment, as the present American public which simultaneously witnesses the tragic consequences of its own vulnerability to the rhetoric of a mythic yesterday, ie, the rise and expected fall of Richard M. Nixon, they will do so for the most part as strangers to one another, celebrating an essential communion only through the inarticulate (and sometimes suspect) public opinion poll.

Second, implicit in phrases such as "culture lag" and "future shock" is the recognition that the human beings, *as they have been fashioned to date*, tend to change at a slower rate than the forms of social life that surround and in some measure define them. It is perhaps that, in their present and past guises, their need for continuity limited their adaptive capacities, though this may be something of a historical variable. We experience this both on an individual and collective basis. Individually, particularly during periods of rapid social change, we often find that the world that significantly shaped our characters and expectations bears little resemblance to the world within which we must function as adults. Moreover, also as an expression of continuing, if not escalating, rates of social change, our children often fail to resemble us in critical details. They fail to resemble us in the same way that we fail to resemble our parents, who may have felt themselves to be to some degree strangers to the world of our childhood. Collectively, we experience this in a multiplicity of versions of the human that may be substantially greater than versions of social orders. In its current state, our global society affords images of clan and tribal warfare, religious warfare and near warfare, racial conflicts, vestiges of class conflict, military cabals, wars of liberation, etc. All of these not only go on within different countries, but in some instances, nearly all forms of conflict are observable within the same country.

Lastly, insofar as the individual experiences nature, the surrounding social world, and in many critical aspects her or his very sense of herself or himself through the mediation of symbol systems, within which the reality of the social is sustained, differences in human response proliferate as there is considerable variation in the ways individuals assimilate and organise such systems. Still greater variations exist in the ways in which such systems on the individual and group levels are transformed by continuing layers of experience. This factor, which may be one of the near constants in the human experience, may not have been important until fairly recently. However, currently, when social consensus is less sustained by a substantial pool of shared or shareable experiences, when behaviour can no longer be efficiently controlled by external observation and constraint, the capacity of symbol systems to achieve greater purity, completeness, and plasticity than is true of the external world they define and implicitly criticise introduces a kind of sociocultural indeterminacy — an indeterminacy that makes the individual-society relationships subject to constant redefinition. Constant redefinition suggests, in turn, a near constant potential for conflict. In a sense, Iago with his capacity for discontent may be closer to universal humanity than Othello with his capacity as a victim.

The inappropriateness of either model — the individual and society in permanent conflict or that of total accommodation — rests primarily in the fact that the conflict never in a very real sense has been between the individual and society, but rather it is one of some people against other people. Society has no needs beyond those which some persons define for it and establish as legitimate. Similarly, society has no basis for constraining individuals beyond the commitment of given persons to given norms, and their ability to achieve legitimate or illegitimate (if these are meaningful terms) control over the means of control. In a final sense, the determination of societal needs or individual responsibilities is not made by wisdom, but by power. The history of all future societies will in all probability not be a history of class struggle. However, this does not mean that it will not continue to be a history of struggle. To talk about the future of individual-society conflicts in its most fundamental aspects is to talk about the future of politics.

Before considering the future of politics in, of necessity, a very general way, it might be appropriate, at this point, to pause briefly and comment upon the politics of the future. In an ultimate sense, little we do or say about the future, at this moment, necessarily gives final shape to the future in any real detail. At best, futurist thinking is a variable of uncertain importance in increasing or diminishing the probability of given outcomes. At the same time, with an erosion of faith in a fixed goal implicit in the historical process — be it global socialism, the second coming of Christ for some, or the Messiah for others — that requires or demands our service, ie, life as a continuing act of existential "bad faith", there is a comparable erosion of mandated imperatives. Conferences on the future tend to have as part of their "hidden agenda" the search for, and legitimation of, just such binding imperatives. The perpetuation of old imperatives (however rehabilitated) or the invention of new imperatives possess a "reality" that is not dependent upon their intrinsic accuracy. We are left to the advocacy of our projections (projections in a double sense). Points of agreement may be of lesser importance than the underlying assumptions about a desired kind of humanity, a desired kind of social order; these also must have their place on the agenda. There are some forms of social life, I would like to think, that most of us would find intolerable however clean its waters, however pure its air, or however prudent its use of what we, with continuing innocence, refer to as "natural" resources.

What then of individual-society and individual-individual conflicts? The most general trend in movement towards a post-industrial, global society appears to produce a paradoxical outcome. Clearly, many of the commitments that presently command (if not create) the passions of people will persist into the future; just as much of the present is disordered by the persistence of values, attachments, or cultural forms of earlier eras. The dominant trends, however, should point in the direction of changes that are congruent not merely with the kinds of almost self-transforming technologies that shape our environments, but more directly by the kinds of transforming experiences they provide. The young Marx asserted before 1848 that capitalism in creating a global division of labour also created a kind of global individual (cf the early sections of *The German Ideology*). History since then appears to represent the gradual realisation of that observation, though we are far from as yet experiencing the full impact of the possibility.

Initially an international division of labour created little more than global inter-dependencies. Increasingly, it creates individuals engaged in activities that occur all over the world and that, in turn, generates identities and commitments that are linked to the activity at the expense of commitments to the more differentiated and specific aspects of the surrounding community or culture. This tendency increases as the organisational structures containing the activities become trans-national in character. This, it must be noted, occurs not only in the wake of the growing number of individuals who act out their specific commitments or activities on a global basis (frequently being sustained by an increasing facility to recreate their personal environments in the midst of many diverse cultural settings — ultimately, all airports, hotels and conference centres look alike), but also is experienced by those who remain physically stationary, while engaged in activities the meaning of which can only be shared by others who are involved in the same activity regardless of the differences, and even traditional conflicts, that otherwsie separate them. The affirming consensus linked individuals and groups who did not share common ancestors, histories, or loyalties.

At the same time, the basis of more traditional loyalties, as they become more abstract and remote from the concrete detail of life, lessen in their ability to command passionate solidarities. A growing capacity for organising a sense of self around personal, rather than social, metaphors creates additionally a capacity for questioning the institutional forms that sustain and are sustained by such social metaphors. The level of questioning becomes pragmatic as it becomes personal. The state, the church, the family command increasingly less loyalty, particularly less unquestioning loyalty, and suddenly find themselves in what is increasingly a "buyer's market", and must learn to market relevance and reward in a manner much like that of merchants of soap and breakfast cereals. They will, to be sure, continue to command a considerable measure of loyalty and, if not loyalty, con-formity on the part of young and old alike,[5] but without the confidence with which they once operated; competence and affluence make it possible to demand a quid pro quo on a personal and immediate level that such institutions have rarely had to face. Commitment is increasingly made on the basis of "limited liability".

The paradox is that as the individual becomes more directly implicated in global systems, that very experience reinforces personal realities over social realities. Many of the clichés associated with horrific accounts of mass society come to mind, except that fear and oppression are seemingly not the only possible outcomes. The process of urbanisation, with its "twin gifts of freedom and loneliness", did not necessarily destroy humanity as such; it transformed it into urban man and woman — a process still incomplete and surely still lamented by many. The creation of global woman and man may surely be no less traumatic, but it is no less possible.

The situation I have described clearly resembles what Durkheim would have termed anomie, a situation where prevailing social institutions and the norms they mandate cease to effectively organise and constrain behaviour: a situation where individuals not only feel free to experiment with the boundaries of the conventional, but also where the guardians of public morality and the conforming public begin to lose faith in their own mission. This situation was viewed by Durkheim as temporary, as collective life would either collapse or achieve re-regulation (the underlying assumption being that the individual was inherently hedonistic and in the absence of effective social control would "degenerate" into something resembling Hobbes' "war of all against all", or the re-establishment of new mechanisms of effective social control). What was inconveivable for Durkheim was the possibility that anomie might become a near-permanent or long-term reality: that the individual's capacity to be self-regulating without being socially dangerous was not only possible, but possibly necessary in order to manage a post-industrial, global society, ie, a world where loyalty is as temporary as specific activities and need not be sustained by intermediary and coercive forms such as race, religion, ethnicity, nationality, or even the family.

Here, I suspect, we might begin to see the major cleavage that will describe the politics of the future. It may turn out to be between those whose concerns for collective social control or regulation (who represent a pressure that might lead to something analogous to what psychologists refer to as "premature identity foreclosure" [or premature future foreclosure]), and those whose very experience requires an exploration of the uncharted terrain of human freedom. A conflict that might well create new and totally bizarre alliances — Nixon, Mao, and Brezhnev facing a common enemy; traditional management and trade unions reinforcing the priority of industrial order; or the ghost of C. B. DeMille editing the films of Antonioni. This commitment may also characterise most of futurism in quest of new, super-personal goals. The conflict is surely implicit in much of current futurist thinking, and in a rather one-sided way, or as the American comic strip figure, "Pogo", once observed: We have met the enemy and he is us.

156

Notes and references

1. The concept of symbiotic relationships in human affairs was probably most significantly introduced into the social sciences with the work of R. E. Park and his students. Paradoxically, the very introduction of the concept of human ecology was also simultaneously the beginning of the end of its utility on either descriptive or predictive levels. Thus, for example, Park and his students could "map" a city like Chicago in terms of about seventy community areas, these areas being "natural areas", ie, unplanned emergences of boundered areas containing specialised activities and/or homogeneous populations. However, the very delineation of these natural areas denied them their natural character, as they became administrative and data-keeping units. In discovery, they were transformed from being parts of the sub-structural to the super-structural, from the ecosystem to the sociocultural system, from the symbiotic to the rationalised.

2. John McHale *The Future of the Future*. (New York, Ballantine, 1971)

3. David Riesman *The Lonely Crowd*. (New Haven, Conn, Yale University Press, 1969)

4. This position allows those who are invested in what is for the moment defined as legitimate to define many forms of deviance from the legitimate, either in the political, intellectual, or personal, as expressions of pathologies; justifying, in turn, carting them off to asylums or other "correctional" institutions. Moreover, the very fact of deviance is then seen as an individual's "fault", and not a failure or deficiency of the existing social order.

5. Many of the old, including most engaged in "futures research", being thoroughly trained to think of themselves largely in terms of social metaphors, tend to panic at the loss of or threat to the "purposes larger than ourselves". We employ dangerously inflated language to deal with what are often limited or tactical problems, feeling more at home on the permanent eve of revolutions, crusades, and impending cultural disasters, where moral worth and a place in history might be established.

The young reflect the fact that we, in the guise of parents and later tutoring ideologues, remain the guides to induction into social life. (Some of us, though, continue to expect the young may possess a more innocent and less corrupt capacity to perceive and act, and, by that fact, form a "children's crusade" that might lead us out of our present cultural wilderness.) They are still instructed in the "distilled" wisdom of history as we continue to believe that the survival of ideas speaks directly to the validity or appropriateness of such ideas. (In that sense, both the contemporary revolutionary and reactionary are conservatives.)

Despite these profound sources of continuity, the very experience of social life calls into question our basic assumptions. Even among those old enough to have invested much of their lives in careers and social movements aimed at goals which, for many, may have lost their meaning *en route* (especially among those who started out to do good but ended by merely doing well), there are visible signs of distress. The young, with less invested, may learn to cut their losses more easily, but that is not guaranteed. Especially as we tend to respond by forcing them into either the empty spaces of social life or encapsulating them in refurbished categories of deviance, applying therapies that produce burn-outs more often than "cures", fearing the one thing we have been trained to fear more than anything else: the freedom to serve ourselves (a temptation to which major segments of industrial societies have been exposed and still others born to). We tend to respond by shoring up commitments to service which are often needlessly protective in character. The persistence of this defensiveness should give way as the cumulative experience of industrial and post-industrial social life both expands and deepens. Nonetheless, it does remain a source for the very cancellation of the future or for severely limiting many of the options for the future as the continuing of this fear makes the present pattern of social life, with all its cruelties and irrationalities, less anxiety-provoking than the uncertainties of new capacities for freedom and equality.

The working groups: a focus on action

Harold A. Linstone

It is a characteristic of most intellectuals that they have a greater affinity for thinking (and writing and talking) than for action. Consequently, a gathering of such individuals rarely produces output other than (a) an exchange of papers, (b) vague resolutions for unrealistic actions by unspecified entities, and (c) plans to hold another meeting. The World Futures Research Conferences are no exception.

In planning the World Special Conference on Futures Research in Rome we felt that a special effort should be made to focus the discussions so that the participants, their organisations, and other groups could bridge the gap between talk and action. We asked ourselves: can the conference suggest priorities for needed research, data gathering, or communication to promote awareness and insight concerning alternative futures?

A document, 'Sketch of Proposals for Follow-up Activities', prepared during the conference planning stage included the following prescriptions:

> The 'World of the Future' has often been marked by the difficulty of directing study to action, for which the need today is becoming more and more urgent. This problem, it has been said, presents even greater difficulties when the action requires group efforts.
>
> For this twofold reason, since its very conception the Special Rome Conference of 1973 — which is intended only as a step forward towards a goal and not as the goal itself — has voiced its purpose of pointing the way to action. It has emphasized this purpose by various means, among them spurring each group to arrive at concrete suggestions of desirable follow-up activities.
>
> As a start we are seeking to collate the ideas received from various participants. They could be similar or completely different from those which will emerge from the separate groups or from the Conference itself. This document has been drafted solely as a stimulus toward a greater effort in this direction and towards the clearest possible output, which could:
>
> 1. lead to clearly identifiable and concrete proposals which permit implementation in the near future;
> 2. avoid duplication with ongoing work;
> 3. be ranked in order of priority based on the consensus of the future research community;
> 4. be open to other and different formulations.
>
> The document also gave some illustration of possible action-oriented proposals, including establishment of a global data bank of forecasts, a project to study practical means of applying holistic approaches to long-range planning of complex systems, a task force to increase awareness of the future (opportunities, threats, alternatives) at the elementary and high school level, and a project to redefine quality of life alternatives.

Nine working groups were formed for the Rome Conference with the following briefs:

1. *Alternative concepts of human development.* Personal development as a complement or alternative to economic development; presentation of images and conceptualisations of persons whose development depends not only on food needs, etc, but also on communication, relationships, etc.

2. *Man-society interactions.* The need for man to accept responsibility for what is required of him by social institutions, and the need for institutions to facilitate the man-society interaction.'

3. *Meta-scenario of alternative quality of life models.* New societies based not only on the feasible but also on the desirable.

4. *Art as an instrument of forecasting.* Art as an indicator of the future and as a means for the development of man's creativity.

5th, 6th and 7th groups: The participation of all people in man's evolutionary transformation; the emergence of diversities and similarities (to be prepared jointly, but developed separately, in the final months preceding the conference and during the conference).

5. *Physical and ecological survival.* Participation of all people.

6. *Self-realisation through communication.* The present state, potential advantages and disadvantages of the 'holistic' approach in the study of systems and the development of communication facilities.

7. *Religion as a human need.* Participation of all people in human development, understood as the integral development of man on the individual and societal level; religion as an integrative factor in human relations; other possibilities of integration.

8. *Alternative social models.* The presentation of such models by countries of the South and the possibilities of common elements with diverse roles and diverse functions.

9. *New technology.* Blueprint for a new technology: old technologies revised, intermediate technologies, innovative technologies, new technological systems, communications technologies.

For these working groups moderators and animators were designated, potential group members suggested and contacted, papers circulated, and a partial dialogue thus begun, months before the actual meeting in September 1973. Then, during the Conference, the groups were each asked to produce one or two proposals they considered to be the most significant outcome of their discussions. All participants were then asked to evaluate each of the proposals in terms of two criteria: *importance* and *practicality.*

The working group schedule called for three sessions. Since there was no closed television monitoring it is impossible to describe and compare their activities. If the typical pattern is followed the initial session was marked by communication problems and wild gyrations in the movement of the discussion. Sometime during the second session these problems were overcome. By the third session the discussions and interactions became productive and the critical element was the rapidly approaching time limit. The ability of the moderator was naturally a very important element in the successful operation of any working group. He would note, for example, that consensus is not essential; a group could well split into two parts and submit one proposal each. In a group of intellectuals the limitation of ego-satisfying flights of rhetoric becomes a major problem. Allotment of sufficient time to develop the most effective wording of the proposal and to prepare for discussion at the plenary discussion is another desideratum which must concern the moderator, also the choice of the spokesman can be a critical factor in effectively communicating a proposal.

1.1 *Education for change

People must be prepared as individuals and small groups for the increasingly rapid rate of sociological and technological change. The individual's ethic must be transformed through an educational process from one dominated by labour to one of inter-personal and intra-personal human fulfilment. This entails a new method of learning stressing inter-relationships with optimal choices. Action should be taken at the local community level and sponsored by central and local governments, as well as private sources (corporations, individuals). Individuals of all age groups should participate in this continuing educational process starting now.

Today the individual is trained by incompatible, anachronistic, encyclopaedic methods which prepare him only for the rigid world of labour. Instead he should be able to adapt himself to, or modify, his situations with self-esteem and pleasure while bearing the insecurity and uncertainty implicit in a rapidly changing world. The concept is considered practical since it operates at the local community level and will not require large funding. Furthermore, participation in the research and demonstrations could educate or re-educate a number of persons fairly quickly.

2.1 Participative conference

Futures researchers are in danger of becoming a self-appointed elite which mirrors the existing elitist system in the society for which they presume to advocate change. A very high priority should be given to the development of techniques and methods which would enable the entire community to participate in futures studies. A conference is proposed which provides the opportunity for simultaneous activities to satisfy many needs and many abilities. It should be organised so that any person can design his own conference agenda and participate simply by collecting information, by exchanging ideas with others or be developing and initiating action programmes. The World Future Studies Federation is an appropriate sponsor. Such a conference could be held as soon as practically feasible and in almost any location in the world.

This concept enables highly informed as well as uninformed individuals to participate informally and formally. It provides for maximum opportunity of choice and freedom as well as structure, and can be adapted to global, regional or local situations. It is practical because it provides for maximum opportunities to participate in planning and implementation. The financial cost is reasonable and does not fall on only one hosting organisation.

---

* The first digit (1) indicates the working group (ie, Alternative concepts of human development), the second is a reference number for the proposal.

2.2 New human rights framework

The unsolved problems of the rights and responsibilities of the individual are compounded by those of organisations and disciplines. This situation calls for change and innovation of organisational and conceptual structures to facilitate man-society interaction. A comprehensive and more sensitive general framework than is currently provided by the Universal Declaration of Human Rights must be created. This process may be facilitated by collecting the existing organised codes of responsibilities or ethical practices produced by professional bodies. The proposal should be submitted by the World Future Studies Federation to the United Nations. This concept entails consultation with international professional and business organisations. Preparation of the framework may take two to five years.

The proposal provides a means for establishing accountability, evolving sensitive educational systems, and for a consensus on the presence of violence and the necessary protections against it.

3.1 Quality of life scenario development

The scenarios drawn up by Western intellectuals for new models of desirable quality of life are inadequate. Scenarios should be built for alternative social futures with full public participation. In addition, mechanisms must be developed which can serve as tools to implement the social innovations suggested by the desirable quality of life scenarios. The scenario development process is expected to take one year (following six months of preparation). Private and semi-public institutions are the likely sponsors and the participants should be experts from selected disciplines and a representative sample of the general public.

The present inflexible and intolerant social and administrative organisations are major obstacles to the improvement of the quality of life. Much preparatory work has already been accomplished in terms of association of needs and social indicators. Judging by the overall dissatisfaction with living conditions, this project should enjoy wide popular support.

3.2 Futures museums

The realisation of quality of life models presents a problem in social mini-technology. Using futures research methodology, the establishment of small-scale, quality of life 'museums' is proposed. The word museum here is used in the original sense (Museion = home of the arts and techniques). Institutions set up for futures research can sponsor and carry out such programmes in cooperation with the citizens concerned.

The activity will facilitate the dialogue between scientists and non-scientists and thereby help to disclose possible and desirable futures for a particular society.

4.1 and 2 Art in action

In general, art can be called upon to (a) aid in the creation of future images; (b) function as a test of the aesthetic quality of the future being suggested; and (c) present and remind us of the irrational (mythical religious, etc) facets of our life. Artistic activity can be the most fertile channel in relating pre-technological and post-technological societies. International groups could be established for the promotion of interactions between advanced technology and the old and new values which still exist in anthropologically diverse societies. Through such studies it is hoped there will be a recovery of values which are declining in the technologically advanced societies and a 'humanising' of technology. The concept is feasible because suitable methodologies exist.

Three special points were emphasised in separate statements by individual group members:

First, an interdisciplinary team should be created comprising anthropologists, sociologists, historians, and artists to 'process' the patrimony of artistic creations to contribute to the understanding of man. It should also study practical ways of removing art objects from the economic sphere to protect unique works, to control art robbery, destruction of archaeological treasures, etc.

Second, there is a need to move art from the concept of a specialised activity. In many primitive societies there are no specialised individuals known as artists – everyone is an artist in his or her own way. In the future we have to strive for a similar diffusion of art, considering that it is a creative image which functions as a part of all activities.

Third, it is meaningless to talk of art as a separate activity. In fact, art education as such a segregated activity should be dropped altogether.

5.1 Lifeboat workshops

Lifeboat workshops are proposed to deal with each of the high priority problems which comprise our present and future ecological predicament. These problems include (a) oceanic pollution, (b) energy resources (both conservation and new sources of energy – particularly solar energy), (c) environmentally oriented development strategies and resource management, especially recycling, (d) excessive urbanisation, particularly alternative experimental communities. These workshops should be sponsored by executive agencies with adequate resources for implementation. The World Future Studies Federation can participate in the generation of ideas; multinational corporations, nation-states, and UN agencies can participate in implementation.

These lifeboat workshops serve as a kind of insurance and help to assure survival with a minimum of suffering in the event of crises materialising in these areas. Feasibility studies have already included confrontation with the user and gained political acceptability.

5.2 Innovation bank

Creation of an innovation bank is proposed to provide seed money, grants, loans, or investments for experiments, pilot studies, and living experiments in those fields of social and technological innovation which are not supportable through normal channels. Sources for funds may be international corporations, governments and world banks.

Alternatives need to be studied well in advance of crises or collapse. At present, the alternatives are too limited, considered too late, and inadequately tested. One reason for this is lack of seed money. More satisfying and tested ways in family living, schools, communications, and democratic decision making may make the difference between hope and despair, survival and collapse. The fund sources mentioned have a self-interest in the long-run survival and well-being of society. Social R & D may be as necessary and valuable as technological R & D was a generation ago. With persuasion and use of a few successful examples, it might become widely accepted with the result that many groups might choose alternatives preferable to present lifestyles.

6.1 Model of the mind

There is a need to know ourselves and to understand much better how our mind operates as a cybernetic mechanism. A model of the mind should be constructed which can show the mechanisms for value determination in a persuasive way. Such a model would aid people in reaching a common basis upon which to discuss attitudes and values. Scientific foundations and research organisations are the appropriate sponsors. This proposal would strengthen the modest level of current work and facilitate important applications of research funding.

This proposal would establish a link between the humanities and engineering. It would help us to understand the difficulties encountered in communication at the inter-personal, group, and cultural levels. The practicality is based on the fact that at least two decades of research experience has already been gained. Steps have been taken already in constructing models and insights have been gained at the elementary school level.

6.2 Futures communication network

At present remotely located groups are unable to confer with each other on vital common issues, eg, villages on farming plans, futurists on world population growth. A task force should undertake the development of a multi-media network which includes progressively cassettes, videotapes, and ultimately a two way real time interactive, computer augmented conferencing network available to all groups in the world interested in the study of problems of the future on a continuous, ie, when wanted, basis. The appropriate sponsor is the World Federation of Futures Studies with UN and/or Foundation support. Full time professionals would be required. The project should begin as soon as possible so that international operation could be effected within four years.

As demonstrated by this conference, communications are not working as well as they should be: remotely located potential participants are absent and communications between successive meetings have been lacking. There are many uses for the same network. The technology is available today and analogous networks are now operating, eg, a worldwide weather reporting system is available to all airports and used in all countries since the 1960s. The proposed concept can in fact use existing networks in many areas at a cost of $10 per hour or less.

7.1 Magic Mountain Centres

Magic Mountain Centres are proposed where 're-creation' of the individual can take place in a holistic way. Such centres will provide space, solitude, skilled teachers, and helpers. They will afford the opportunity to commune with oneself, one's family, and others of interest. The leaders of Magic Mountain must be teachers and gurus who do not control but facilitate and bear witness by their own behaviour to the processes required. Those who come to Magic Mountain can come for short or long periods. Resources of art, religion, and science will be available to help people move between the known and the unknown in new and open ways. The sponsor suggested is the Committee of Correspondence on Religion of the Future which is being set up through this conference (see below). Initially, centres in Europe, Asia, Africa, North and South America are advocated.

Futurists and others need to learn to work in new ways in order to be able to create futures, not only theorise about them. Ideas and starting points for this kind of concept already exist in France, the UK, Japan, North America, and other places.

7.2 Religion communication network

There is a strong need for a continuing mechanism for conference participants and others to study and work with religion as a human need and as a liberating, developing, and integrating factor in human relationships. Innovative networks should be created for thinking and working in new ways upon issues of futures creation that are related to religion. Examples of projects are the comparative study of basic religious symbol systems and their future creating potential, global modelling based on Buddhism and Hinduism as well as Western values, exploration of relationships between art, science, and religion, and development of innovative centres such as the Magic Mountain proposal. This activity is to be begun immediately through the Rome Office of IRADES but with local research and development projects in any location.

8.1 Societal alternatives in schools

To effect a change of direction, it is vital that our existing value system of society be modified to accept alternatives. Basic educational institutions should include in their curricular, courses on different cultures, human futures, alternatives, and social models and values. The first condition is to change the philosophical basis for the egocentric, atomistic, competitive, uniformistic, hierarchical ways of thinking into attitudes which stress consideration for others, inter-relationships, cooperation and symbiosis, participatory decision making and awareness of the existence of diversity and recognition of its positive contribution. Homogeneous majority rule should be changed to permit diversity; economic inequality should be altered to obtain equality through collective or communal ownership. Production and distribution should take into account the differences of skills and needs of people. The institutional change should be made to allow the freedom of opinions and expression, with no control or censorship. The key to this proposal is the educator and communicator at every level, particularly in the industrialised countries.

Alternatives for the future require radical departure from the value system that wasteful affluent societies have developed for themselves. The existing educational and communication systems can be used to transmit the proposed new direction with a minimum of additional financial obligations. It should be noted, however, that the existing structured educational systems must also undergo change.

8.2 Restructuring of decision making

The existing power structure protects the exploiters and oppresses the weak and deprived. Decision-making processes should be reversed and decisions should come from the base to the top. The restructuring should be founded upon the principle of self-government, heterogeneity, and constructive use of diversity. This effort should begin within the next five years and should involve those who are both inside and outside the power structure and are aware of the necessity for restructuring.

Those without power could be helped by a fundamental restructuring of societies within the power structure of the world. Many individuals share awareness of this need, but do not know each other and therefore do not communicate. A communication network would be of obvious use to them.

9.1 Technology Centres

Human needs are not being satisfactorily met by the technological system as it has evolved, whether under capitalist, socialist, or any other auspices. The further development of technology along the dominant established lines does not promise solutions to the most serious problems now confronting mankind, such as exhaustion of resources, pollution, low quality of life, alienation, unemployment. There are gaps of knowledge and, even more important, gaps in the availability and application of knowledge, which will not be filled without some new effort. Technology centres need to be created, more orientated towards human needs in both developed and developing countries. Their task would be to ascertain, systematise, and disseminate and, where necessary, encourage creation of relevant technological knowledge by practical experience, as well as systems to supply better technology so that people everywhere could obtain help in solving their problems. The centres should ensure that there is a continuous process of experimentation in the search for solutions to the great recognised problems alluded to above, as well as potential problems.

Several recurring themes are evident in these proposals:

(a)   Communications networks (proposed not only by the Communication Group (6.2) but by the Religion Group (7.2) and in some degree by the Man-Society Interaction Group (2.1) and the Alternative Social Models Group (8.2)).

(b)   Creation of centres to attack critical socio-technological problems (proposed by the New Technology Group (9.1) and the Physical and Ecological Survival Group (5.1 and 5.2)).

(c)   Creation of centres to permit more holistic insight on alternative futures (museums proposed by the QOL Scenario Group (3.2) and the Magic Mountain Centres proposed by the Religion Group (7.1)).

The evaluation

Before reviewing the results of collective evaluations of these proposals by the Conference, I would like to add a few personal impressions concerning the plenary session itself.

There has been much talk among futurists about participative planning. The

Table 1. PARTICIPANTS' EVALUATION OF PROPOSALS

Percentage distribution of ratings of proposals by participants

	Importance			Practicability		
Proposal description	A	B	C	A	B	C
1.1 Education for change	60	30	10	10	43	47
2.1 Participative conference	24	40	36	30	44	26
*2.2 New human rights framework	46	26	28	42	28	30
3.1 QOL scenario development	53	38	9	31	40	29
3.2 Futures museums	24	49	27	40	38	22
4.1 and 2 Art in action	23	31	46	23	31	46
†5.1 Lifeboat workshops	69	20	11	34	52	14
†5.2 Innovation bank	61	21	18	41	31	28
6.1 Model of the mind	24	40	36	23	32	45
†*6.2 Futures communication network	42	39	19	47	39	14
7.1 Magic Mountain centres	35	32	33	40	33	27
7.2 Religion communication network	28	32	40	49	35	16
8.1 Societal alternatives in schools	70	23	7	24	36	40
8.2 Restructuring of decision making	52	21	27	7	24	69
9.2 Technology centres	38	43	19	22	58	21

*High importance and high practicability by first method
†High importance and high practicability by second method

Importance
A critical, urgent, highly desirable
B important, useful, desirable
C secondary, unimportant, undesirable

Practicability
A implementation offers no major problem, sponsor organisation can be 'sold', capability exists to perform task and can be marshalled, cost reasonable
B implementation offers some major but not insuperable problems, sponsor may be obtainable if strong effort is made, capability to perform activity partially exists and gaps can be filled, costs high but bearable
C unrealistic, little support can be found, costs beyond reach, capability to perform task does not exist

subject is directly mentioned even in the proposals (2.1 and 8.2). Implementing the concept is apparently difficult as the final plenary session itself illustrated: less than half of the attendees made any oral contribution. It was ironic to observe that future-oriented groups resorted to the most conventional, old-fashioned means of getting their message across and 'marketing' their ideas. Innovation in presentation was notably absent. Those groups giving at least some thought to the matter chose the most effective communicator in their group. His or her success depended inevitably on the proposal itself. The more concrete and limited in scope, the easier the task. The more vague and global the proposal, the harder it was to gain acceptance in terms of the criteria of evaluation. Responses to the question 'Who should do it?' or 'When and where should it be done?', eg,

'Everywhere and at every level' (8.1)
'Everywhere and continuously' (8.2)
'Individuals at all age groups (in) home, schools, community milieus' (1.1)

made the spokesman's assignment exceedingly difficult. Considering the stipulated working group concept, the non-specific and unrealistic nature of many of the proposals presented must be viewed as the major disappointment of the entire effort.

Table 1 presents the results of the working group participants' evaluation of the action proposals. Let us consider the first of the two criteria — Importance. There were four proposals which were rated A in Importance by at least 60% of the respondents: 8.1, 5.1, 5.2 and 1.1. We now ask the question: Which of these proposals of utmost importance are also considered practical or feasible? We find that only one, Proposal 5.2, has at least 40% rating of A in that category.

Lowering our sights somewhat, we consider next the proposals which received a 40 to 60% rating of A in Importance. These are 3.1, 8.2, 2.2, and 6.2. Again, we move over to the Practicability criterion and we find that of these four, only two, Proposal 2.2 and 6.2 gained at least a 40% rating of A in Practicability.

This particular procedure thus points to three proposals as deserving particular attention, ie, promising combinations of both high importance and high practicality: 2.2, 5.2, and 6.2. One alternative procedure is simply to multiply the A percentages for each proposal. If this is done the highest-ranking proposals are 5.1, 5.2 and 6.2. There are obviously many other ways to evaluate proposals such as these and the reader is encouraged to do his own. The purpose of this exercise has been to focus and crystallise the wide spectrum of discussion and to provide at least initial guidance for follow-on work.

Table 1 deserves a few additional comments. Notice that Proposal 8.1 is considered of high importance by more people than any other (70%) yet it is at the same time considered relatively impractical. Similarly, Proposals 1.1 and 8.2 received very high importance and very low practicability ratings. Group 4 appeared to have considerable trouble generating proposals and agreeing upon a presentation. This is reflected in the Art Group 4 input, which is ranked C in both areas by nearly half the attendees. Proposal 6.1 also fared quite poorly in both categories.

Conclusions

Three points merit particular emphasis. First, effective *communication* is vital in all aspects of human activity, but is a particular challenge in futures research. You may have the greatest ideas but if you cannot get them across to others, they are wasted. I think several of the proposals that did not rank high were excellent and deserve much attention, but determination of the most suitable means of communicating them to others is essential. If we cannot communicate well and imaginatively amongst ourselves, how can we possibly expect to communicate with others in different disciplines, cultures, and activities?

It is evident that many futurists still cannot bring themselves easily to think in concrete terms. Nevertheless, a number of intriguing and practical proposals were uncovered. The newly created World Future Studies Federation could well take its cue from suggestions such as those on the creation of communication networks and centres to study intensively critical socio-technological problems. IRADES has expressed interest in the Religion Group's Network (7.2). There is clearly a need for insights on alternative futures which is not being met at this time. There are specific tasks which can be – and certainly should be – promoted and undertaken by responsible futurists. Undoubtedly many of the Working Group members would be pleased to participate further in crystallising and refining their proposals.

Finally, I am convinced that the Working Group concept has a definite role in interactive, transdisciplinary futures research. Although the duration of the exercise was too brief, it clearly suggests that a synergistic effect can be achieved through such a mechanism. And this presents the strongest argument for its use. We have limited time and brainpower to overcome crises and convert them into opportunities, and at this juncture of futures research, coordinated work in small groups is more desperately needed than expansive talk in large sessions.

Futures studies-quo vadis?

Yehezkel Dror

The IRADES Conference Human Futures: Needs, Societies, Technologies *provided a convenient, appropriate and much-needed opportunity for evaluating the present state of futures studies and identifying some guidelines for their futures: where can and where should futures studies go and what changes are needed for optimum development?*

Futures studies have reached an appropriate point for self-examination and self-direction, and this Conference provided an excellent occasion for doing so, for two reasons: the critical stage reached by futures studies and the character of the Conference.

Started by Bertrand de Jouvenel about 13 years ago, futures studies cannot claim any more to be an infant whose every weakness can be expected to pass away on its own with growing age. Many of the external signs of maturation already characterise futures studies: five international conferences and many smaller formal and informal gatherings; courses at universities; professional periodicals in several languages; books devoted to methodology and contents; special research institutes; and local and international professional and semi-professional associations – all these exist and are proliferating. Whether such numerical and nominal indicators of growth really signify qualitative maturation as a serious intellectual and human endeavour – this is a different question. But, at least, futures studies is by now a sizeable phenomenon that can be identified, even if its borders are somewhat fuzzy; that can be diagnosed; and for which prescriptions for improvement can be proposed. Indeed, it is the rapid growth of futures studies activities which makes urgent a critical and improvement-oriented examination, so that congenital defects and malignant trends can be corrected – if necessary, by radical surgery – before critical hypertrophy sets in and futures studies as a whole become doomed.

The IRADES Conference provided a nearly perfect opportunity for considering the futures of futures studies, which permitted futures studies to appear at their best:[1]

- The audience was small enough to permit meaningful interaction, sufficiently select to keep out many of the more ignorant self-styled 'futurists' who plague futures studies, large enough to reflect[2] different cultures and styles, and sufficiently diversified to permit fruitful interaction and cross-stimulation.
- The subject of the Conference was clearly defined to provide a focus, while also broad enough not to become a straitjacket repressing creativity and originality,

so far as these were present.

- The Conference was carefully structured, with a mixture of prepared addresses, plenary discussions, work groups and informal meetings — in a way conducive to accumulative multidisciplinary work.
- While open-ended, the Conference was unique in its attempt to process some of the outputs in the form of operable implications. The schema developed by Harold Linstone provided an appropriate tool for trying to sum up and test some of the results,[3]
- Last, but not least, the residential basis of the Conference, the contemplation-encouraging physical environment and the excellent facilities provided by IRADES reduced noise and external friction to the minimum, permitting full devotion of all energy and capacity to the substantive work of the Conference.

These features provided an optimal environment for futures studies and therefore, all the weaknesses displayed at the Conference must be viewed as genuine features of the subject as a whole (though certainly not of each and every futures scholar and futures research item), undistorted by external constraints and dysfunctional conditions. Similarly, the points of strength demonstrated at the Conference can also be regarded as integral characteristics on which the progress of futures studies can be based.

The Conference, as already mentioned, glaringly showed off some of the main weaknesses of futures studies, such as: sloganism, dogmatism, repetition of the obvious, fixation on pet solutions, conservatism even in radicalism, amateurism bordering on ignorance,[4] wishful thinking and tunnel-vision. Usually hidden behind high-sounding phrases, these and other weaknesses were clearly exposed by Linstone's exercise, which forced participants to transform verbal symbols into operational formats.[5] But I do not wish to delve this time into such weaknesses.[6] Neither do I want here to enumerate the equally important hopeful features of futures studies demonstrated at the Conference, which are well reflected in the papers in this volume. Rather, trying to follow a positive approach, let me suggest six main features of 'good futures studies', which can serve both for balanced evaluation of the present state of futures studies and, more important, as a guide for their advancement and development.

Six features of good futures studies

(1) *Value-sensitivity explication and analysis.* Futures studies are very value-sensitive. Findings, analyses and conclusions are very dependent on subjective values, even in the domain of so-called 'objective' forecasting or behavioural prediction, where we try to identify possible and probable futures, as independently as possible from our preferences, fears and values. The weakness of predictive methods and the importance of subjective estimates even within strict methodologies[7] reinforce the influence of values in futures studies, much more so than is the case in 'normal' sciences. Certainly, the role of values becomes dominant when we enter the domains of normative predictions, which are legitimately and explicitly highly dependent on values.

This dependence introduces a very subjective element. For some purposes, it may be regarded as legitimate and even desirable; for others, it is an unavoidable evil. But, in all cases, the high value-sensitivity of futures studies must be recognised, explicated and analysed. Various futures scholars do and should differ in their value preferences. This applies even more so to the clients of futures studies, including politicians who, after all, are the legitimate value-judges according to all political ideologies.[8] To reach secondary agreement among futures students – that is, agreement on what one disagrees about – to avoid usurpation by futures scholars of functions that do not belong to them, to prevent futures studies from becoming another dead hand of the past trying to control the future,[9] and to increase 'positive redundancy' within futures studies, the value-sensitivity of futures studies must be explicated and then analysed with all the emerging tools of value-analysis and value-invention.

This desideratum of good futures studies will be loathsome to persons with little tolerance for value-diversity; and, even more so, to those fellow travellers of futures studies, who want to use the latter as a vehicle to push their own dogmas. But I think there can be no doubt that value-sensitivity explication and analysis are absolutely essential for the maturation of futures studies.

(2) *Creativity and imagination.* There will be less disagreement on my second characterisation of preferable futures studies as creative and imaginative. In fact, the concepts of 'creativity' and 'imagination' are often used as a battle-cry by those who support a radical *laissez-faire* ideology for futures studies, with every opinion being entitled to the same respect, never mind if based on years of study or incidental day-dreaming.

Because of this appearance of consensus, I think that the requirement of 'creativity and imagination' needs careful consideration. At least, the following points must be emphasised:

- 'Creativity and imagination' have some discipline of their own, such as internal consistency, value-explication and consideration of feasibility.
- 'Creativity and imagination' in the context of futures studies must be substantive in contents, with new values, new social institutions and new social laws being elaborated and discussed – rather than most energy going to pure exhortation for others to be creative and imaginative. Typical are statements such as 'we must find new values to solve the problems of the third world' or 'a new kind of humane technology must be invented'. A few such need-specifications are acceptable. But when such demands exhaust the creative and imaginative contents of most papers and conferences – then this is very bad.
- Creativity and imagination are urgently needed not only on where we should want to go, but how to get there. Ideal futures should be designed; but without some feasible scenarios on how to move in their direction, little will be achieved. Hence the need to search for social inventions that are operational now, rather than nearly exclusive concern with far-off futures into which to escape mentally.
- The scarcity of creativity and imagination must be recognised, with resulting differentiation between the respect to which different items of creativity and imagination are entitled. In other words: everyone claiming to be creative and

imaginative should be given a chance or two. But there is nothing in democratic ideology, nor in professional ethics which demands that limited time and overburdened attention should be equally allocated to all who declare themselves again and again to be 'creative and imaginative' and who are continually re-inventing a broken wheel.

These points are designed to stimulate and encourage real creativity and imagination about the future, which are in danger of distortion. Radical creativity and imagination are urgently needed in futures studies and are sorely missing from them – in part, I think, because of the utilisation of the terms of 'creativity and imagination' as a cover-up slogan rather than as a challenge for hard efforts.

(3) *Improving methodology* is, without doubt, a main requirement of futures studies as a professional and science-related activity. Futures studies are broad in sense and aim, going beyond the boundaries of a scientific discipline. Therefore, the methodologies of futures studies should include apperception, serendipity and even carefully controlled altered states of consciousness. But the need for innovative and even daring methodologies is no excuse for methodological laxity. The opposite is true: very advanced methodological sophistication is particularly needed, exactly because of the reliance of futures studies on a broad range of novel methods, methodologies and tools.

In this respect, the following methodological needs deserve special emphasis:

- Necessity for strict differentiation between values, assumptions, dogmas, facts with different degrees of validation, predictions of varying reliabilities and a range of types of conclusions and findings.
- A probabilistic view of the future as a basis for study with some deterministic aspects on one side and some arbitrary elements on the other. This introduces into research all the methods of analysing and dealing with probabilistic phenomena.
- Need for a comprehensive systems frame, within which the interaction and interdependencies of various futures of discrete aspects and institutions is carefully examined and correlated – such as through the tool of 'cross-impact analysis' and the concept of 'alternative comprehensive futures'.
- Combination of rational and extra-rational tools, elements and components, within defined methods.[10] To this should be added the consideration and even utilisation of irrationality as a factor in and for futures research.
- Some distinction between futures studies as an intellectual activity on one hand and social prophecy, contemplative dreaming and futures-oriented political action on the other hand. These interact, but constitute different activities.
- Constant self-doubting, self-searching and self-evaluation – as an essential approach to improving futures studies' methodologies through feed-back and iteration.

Probably, futures studies should have a broader methodological base, like policy sciences.[11] In any case, improving the methodology is a most critical element of good futures studies.

(4) *Multidimensional and combinational.* Multidimensional means bringing together different disciplines, methods, methodologies, approaches, values, appreciative frames, cultural backgrounds, personal life experiences and personality types. It also implies a broad range of alternative assumptions as bases for all facets of futures studies. 'Combinational' means activation and application of the different dimensions with their various sub-components together, but without complete fusion. Interaction between varied elements from different dimensions is required to enrich futures studies and improve their validity through positive redundancy. But complete integration is impossible because of lack of a unitary basis; over-integration at the present state of futures studies may also increase the dangers of shared bias, common mistakes and identical blind spots.

(5) *Clinical attitude together with deep human concern.* Futures studies presume to become an important part of the most ambitious of all of human endeavours, namely the attempt to influence the future through conscious collective and individual action. To make a useful and unique contribution, futures studies must adopt a clinical approach to the realities and potentials of so-called *homo sapiens* (including the possibilities of altering the species). But, dealing with the fate of conscious beings, deep personal concern with the feelings of individuals must underlie the clinical approach. This dichotomy brings unavoidable tensions, which should serve as a stimulant. Letting spiritual concerns dominate will empty futures studies of their intellectual contents and nullify their utility as a conscious and responsible guide to human action; letting clinical attitudes displace human concerns will sterilise futures studies morally and nullify their significance for the more fundamental issues of shaping the future.

(6) *Diverse outputs.* Futures studies outputs differ. They include, among others: (a) knowledge, as itself a main human aspiration; (b) inputs into policy making; and (c) inputs into individuals, whether scholars or laymen, intellectuals or 'feelers' – the world-view and self-view of all of whom is enriched by the broad perspectives supplied by good futures studies.

Six commandments

These six features of preferable futures studies present my subjective opinion. The reader may wish to revise these characteristics. But a consistent, dynamic and explicit image of good futures studies, to be itself revised constantly, is necessary for evaluating the present state of futures studies and for guiding its advancement. There can be different legitimate opinions on the desirable features. But to let futures studies drift on without critical self-examination and without explicit, open-ended and dynamic guidance – this, I think, is more and more unacceptable.

A possible next step is, to take up these six characteristics of preferable futures studies and use them as criteria for evaluating the present state of futures studies and their probable future, if they continue to develop along their present lines.

Let me proceed by presenting six positive recommendations which I regard as operational and feasible, and as useful in the sense of helping to move futures studies in the direction outlined above:

My first recommendation may look somewhat mundane and perhaps even a little brutal: I think every person coming to a conference on futures studies should take first the trouble to read, at least, ten books[1,2] on futures studies methodology. This is essential, so that some shared concept package exists, on which common work can be based. I would be ashamed to make such a recommendation, were it not for the clear fact that even at so carefully organised a meeting as the IRADES Conference, more than half the participants demonstrated, in my perception, amazing ignorance in basic futures studies knowledge. This is an intolerable situation – hence this recommendation.

Second, I think the time has come to pay more attention to developing some ideas in depth. Workshops of one to two weeks, where a small group of carefully selected, but diverse, participants work intensely on well-defined subjects – may be much more useful than additional global conferences. Also urgently needed are more research centres and intensified work by existing research centres. Work in depth – this, surely, is a main need of futures studies.

Third, I think more thought should be given to the desired outputs of futures studies, so that specific activities can be oriented towards explicit goals. Outputs, as already noted, can and should be multiple. But they should be specified, as a guide to activities and so as to permit learning from feed-back. At present, quite a lot of what goes on in futures studies seems to serve mainly cathartic needs of individuals. Maybe this is legitimate and cheaper than psychoanalysis. But I think many of us get tired of this phenomenon, which may drive out those of us who regard futures studies as oriented towards some results, intellectual or otherwise. 'Output' does not necessarily mean something measurable. Educating ourselves, satisfying human curiosity about the future – these are also legitimate outputs. But we should specify the main outputs towards which to orient futures studies, so as better to structure futures studies activities accordingly.

The fourth recommendation, which applies to all of us, is: strict self-discipline! The absence of clear-cut and accepted quality criteria and of a homogenous group of peers which exercises judgment – are an advantage of futures studies, keeping them more open and providing a chance for rapid innovation. But they also reinforce the dangers of bad work. Each one of us must, therefore, exercise strict self-discipline as a part-substitute for the absence of outside quality control. 'Is what I think a slogan, is it a declaration, is it perhaps important but repetitious, what is its base, is it validated?' – these are among the questions that all of us should pose to ourselves when looking into the mirror of self-searching criticism.

Fifth, we must combine an open mind with selectivity. If over-selective, we will miss important cues; if too open, verbiage will drag wisdom down the drain of ignorance.

Sixth and last: I think we must increase the number of basic assumptions which serve as bases for our thinking on futures. We cannot assume that human beings are mainly good or bad, that wars will/will not happen etc.[13] More diversity in basic assumptions is essential to broaden our thinking enrich our perspectives and increase the probability that futures studies may fit and serve future realities.

Until now, futures studies engaged mainly in self-recruitment, building up of some self-identity and construction of an initial international network. Also, futures studies engaged in problem identification and in selective penetration into some issues. I think that from now on we must try to be more comprehensive, more sophisticated and, especially, go deeper. This depends on persistent and intensive work. Travelling from one congress to another or writing a paper once a year and, in the meantime, doing other things — is not sufficient for advancing future studies. Professional commitment on a sustained basis is essential.

As already mentioned, futures studies is a part — not the only part, but an important part — of the most ambitious of all human endeavours, namely the attempt, by conscious action, to influence the future. There exists no more ambitious aspiration than to shape the future through conscious, social action. To have some chance of success in contributing to this endeavour, we futures scholars must work very hard. Also, I think, we need a kind of philosophic attitude, combining enthusiasm, wisdom and stoicism. If we try very hard, we may succeed a little. But, in order to succeed a little, futures studies must mature.

References

1. May I use this opportunity to express my appreciation for this Conference as sponsored and organised by IRADES and especially by Don Pietro Pace and Dr Eleanora Masini.
2. Some action-oriented futurists misuse the term 'representative' in this context, as if any futures scholar or conference participant has a mandate to 'represent' anyone, but himself. I prefer the term 'reflective' to indicate one of the desired features of futures studies, namely to reflect diverse cultures, points of view and approaches — as long as this does not impair the main qualities of futures studies.
3. See preceding paper by Harold Linstone.
4. Let me try to illustrate what I mean by this list of accusations, by picking up this point: after reading the papers submitted to the Conference, I selected 10 of them which proposed, as major innovations invented by the authors, a number of ideas fully covered in literature for quite some time. Then, at the Conference, I delicately brought up the matter with some of the authors who were present. The response was similar in content though different in formulation: the authors had not troubled themselves to read futures studies and related literature. This is bad enough; but some of the authors added arrogance to ignorance, presenting themselves as so original as to make reading of literature unnecessary for them. At any professional conference, papers clearly based on ignorance of available knowledge by the authors would be severely criticised. At futures conferences, authors were amazed that I dared ask them whether perhaps they are familiar with this or that main item in futures studies literature! At none of the meetings were such papers attacked as ignorant, either because of misplaced politeness, or because most of the audience too does not know the literature, or because good intentions are accepted as a full compensation for intellectual laziness — I dare not guess which.
5. I do not wish to imply that the only or even the main test of good futures studies is 'operational implications'. Robert Jungk was right when pointing out that other outputs are also important, such as self-enrichment, mutual stimulation and longer-range educational effects. But surely, capacity to be transformed into operational recommendations, at least for further research and studies — is a very significant standard for evaluating the quality of futures studies.
6. For an attempt in that direction following the Bucharest Third World Futures Research Congress, see Yehezkel Dror, 'A Third Look at Futures Studies', *Technological Forecasting and Social Change*, Vol. 5, No. 2 (1973), pp 109-112.
7. The importance of 'subjective probabilities' in many strict forecasting methods and the dependence of the widely hailed 'Delphi Method' on intuitive feelings and, at best, tacit

knowledge – serve to demonstrate the dependence of predictive methodology as a whole on factors which are highly sensitive to values.

8. Ideologies differ in the identity of 'legitimate politicians', from self-appointed 'Führers' to elected officials and even the population at large. But all ideologies agree that value-judgment belongs to the legitimate politicians and not to any group of intellectuals or professionals, such as 'futurists', unless we want the latter to become a new ruling elite, ie, 'legitimate politicians'.

9. Very apt is the phrase coined by Johan Galtung, that futures studies may become a form of 'colonialism by the present over the future'.

10. On the parallel problem of combination between rationality and non-rationality within the context of policymaking, see Yehezkel Dror, *Public Policymaking Reexamined* (N.Y.: Intext Educational Pub., 1968 and Aylesbury, Bucks: Leonard Hill Books, 1973).

11. For a discussion of futures studies from the perspective of policy sciences, see Yehezkel Dror, *Ventures in Policy Sciences* (N.Y.: American Elsevier and Amsterdam: Elsevier, 1971), chapter 5.

12. I picked the number 'ten' arbitrarily, probably because of emotional affinity to the Ten Laws of Moses. Any other number large enough to provide familiarity with main concepts, methodologies and substantive ideas of futures studies, will do. After a number of books are absorbed, additional readings tend to rapidly decreasing marginal returns – because the total stock of futures studies knowledge and ideas is still quite small.

13. For an attempt to move in this direction with the help of prototypes of low-probability extreme situations, see Yehezkel Dror, *Crazy States* (Lexington, Mass: Heath Lexington, 1971).

Authors

John McHale is Director of the Center for Integrative Studies in the School of Advanced Technology, State University of New York. He has a PhD in Sociology and has published extensively on the impact of technology on culture, mass communications, and the future. His latest books are *The Future of the Future, The Ecological Context* and *World Facts and Trends*. Dr McHale is also an artist and designer, and his work includes graphics, exhibition design, television, film, and general consultancy to organisations in the USA and Europe. He is a Fellow of the World Academy of Art and Science, the Royal Society of Arts (UK) and the New York Academy of Sciences. He was awarded the Medaille d'Honneur en Vermeil, Société d'Encouragement au Progrès in 1966. He is a member of learned societies including the World Future Society and the Continuing Committee of the World Future Research Conferences, and is a member of the *Futures* advisory board.

Maurice Guernier is a Doctor in Economic and Political Science, University of Paris. He holds a post in the Civil Service of France, where he is inspector of the Ministry of Finance and Economic Affairs. He is a member of the staff of Jean Monnet in the Commissariat Général au Plan; he was also economic counsellor to the Prime Minister. From 1948, Mr Guernier was in charge of development research for governments in Africa and the Near East, not only in global development but also in rural schemes as in Chad, Dahomey, Rwanda and Burundi and Cameroun. He is at present a member of the Club of Rome, specialising in the futures of the Third World.

Lewis Mumford studied at the public school of New York, at the City College of New York, Columbia University, and the New School for Social Research. He has an Hon LID, from Edinburgh, and is Hon Dr Architecture from Rome. He is a pioneer in American studies, urban studies, the history of technics, and in "social ecology". He has held many academic posts in American universities and was co-chairman of the Conference on Man's Role in Changing the Face of the Earth, Princeton. His principal books include *Technics and Civilization, The Culture of Cities, The Condition of Man, Art and Technics, The Transformation of Man, The City in History, The Myth of the Machine, Vol. I, Technics and Human Development, Vol. II, The Pentagon of Power.* He is a member of many learned societies and has won numerous honours and medals.

Sam Cole is a member of the Science Policy Research Unit, University of Sussex, UK. He studied theoretical physics at Imperial College, London, Sussex University and Cambridge University. He has worked on urban planning research for the Department of the Environment, which involved studies of social indicators and quality of urban life styles. He returned to Sussex to join the research group STAFF – Social and Technological Alternatives for the Future. Mr Cole was one of the authors in *"The Limits to Growth* controversy" (*Futures* Special Issues, Volume 5, Numbers 1 and 2) where he described and evaluated the world models used. He presented the SPRU critique at a meeting of the Club of Rome.

T. Craig Sinclair is at present Head, Department of Innovation Management, International Institute for the Management of Technology, Milan. After graduating in physics at Edinburgh he worked in medical research and development and in the public and private nuclear power industries, during which time he wrote *Control of Hazards in Nuclear Reactors.* He then worked at the Science Policy Research Unit at the University of Sussex and published books on innovation and safety. *"The Limits to Growth* controversy" *(Futures,* Special Issues, Volume 5, Numbers 1 and 2) occupied the remainder of his stay at SPRU. He is co-editor of *Research Policy,* a journal which is devoted to research management and planning, and he has recently edited *Environmental Pollution Control.* He has acted as consultant on environmental and risk policies to the UK government, OECD, UN and IAEA, and is currently concerned with the effects on innovation of institutional control procedures.

James Allen Dator is Professor of Political Science, and head of the Program in Futures Research of the Social Science Research Institute of the University of Hawaii. He obtained degrees from Stetson University (BA), the University of Pennsylvania (MA) and the American University (PhD). He has also done post-doctoral study at various American universities, and taught at Virginia Polytechnic Institute, Rikkyo University (Tokyo, Japan, for six years), and the University of Maryland. He is advisor to the Hawaii State Commission on the Year 2000. His TV course, "Tune to the Future", won an award for creativity from the National University Extension Association. Professor Dator is active in the World Future Society and is a member of the Continuing Committee of the World Future Research Conferences. He contributed to the Conferences in Kyoto and Bucharest, and to an international group of futurists in Rome in 1971. He has published three books, co-edited another and has published numerous articles.

Harold A. Linstone has a BSc from the City College of New York, a Master's Degree from Columbia University, and a PhD in Mathematics from the University of Southern California. He is Professor of Systems Science at Portland State University and Director of its new PhD Programme. He is senior editor of the international journal, *Technological Forecasting and Social Change*. He has served as a senior member of the Rand Corporation, and has introduced a course at the University of Southern California, "Planning Alternative Futures" which won him an award for Innovation in Teaching. He has presented his views on future needs and capabilities to a subcommittee of the US Congress and has given seminars on Technological Forecasting and Long Range Planning in the USA, Germany, the UK, France, and Israel. He is presently working (with Murray Turoff) on a book about Delphi.

William Simon studied at the University of Michigan, Wayne State University and the University of Chicago, where he obtained his PhD. At present he is Program Supervisor in Sociology and Anthropology at the Institute for Juvenile Research in Chicago. He has held lectureships in sociology at Roosevelt University and Southern Illinois University and was Assistant Professor in the Department of Sociology at Indiana University. He is associate editor of *Social Problems* and a member of the editorial board of *Archives of Sexual Behaviour* and *Journal of Youth and Adolescence*. He is a prolific writer of articles and professional papers on social behaviour.

Yehezkel Dror is Professor of Political Science at the Hebrew University of Jerusalem. His main areas of research are policy making, strategic analysis and policy sciences – subjects in which he has published seven books and numerous articles. Professor Dror has served as consultant to various governments and international organisations, was a Fellow at the Center for Advanced Study in the Behavioural Sciences (Palo Alto), worked for two years as a senior staff member with the Rand Corporation (Santa Monica and New York) and has received Israeli and other awards recognising his scientific and applied contributions. Professor Dror is President of the Israeli Futures Research Association, a member of the advisory board of *Futures* and a member of the Continuing Committee for World Future Research Conferences.

IRADES Istituto Ricerche Applicate Documentazione e Studi is an independent, non-profit institute for applied research, documentation and studies for human and social forecasting. President: Hon. Flaminio Piccoli; Secretary General: Pietro Pace. Address: via G. Paisiello 6, 00198, Italy. The Institute is concerned to inform policy makers and those whose responsibilities are future oriented. The areas covered are the family, education, social communications and religion, and IRADES aims to provide solutions to problems arising from change. Periodical publications: *Social and Human Forecasting* (Newsletter), *Social Forecasting, Abstracts* (Bulletin) and *Social Forecasting Documentation* (Annual).

FUTURES was launched as an international, quarterly journal in September 1968 following the First World Futures Conference in 1967 in Oslo. Co-operation with the Institute for the Future, Menlo Park, California, USA began in June 1969, and the frequency of publication of FUTURES was increased to six times a year at the beginning of 1973 to cope with the growing interest in futures studies throughout the world. The editorial coverage spans the developing methods and practice of long-term forecasting for decision making on the future of man, society, economy, technology and politics.

FUTURES' editors are John Thomas, Ivan Klimeš and Elizabeth Teague with Henry David acting for the Institute for the Future, and its advisors are Daniel Bell, Gabriel Bouladon, James Bright, Robin Clarke, Yehezkel Dror, Emilio Fontela, Peter Hall, Olaf Helmer, Christopher Jones, Robert Jungk, Herman Kahn, Hidetoshi Kato, John McHale, Hasan Ozbekhan, Keith Pavitt and Geoffrey Vickers.

FUTURES, IPC Science and Technology Press Ltd, IPC House, 32 High Street, Guildford, Surrey, GU1 3EW, UK.

Presidential Committee

Flaminio Piccoli — Italy
(President)
Bertrand de Jouvenel — France
Thomas Adeoye Lambo — Senegal

Scientific Council

Giorgio Nebbia — Italy
(President)
Eleonora Barbieri Masini — Italy
(Executive Secretary)
Francesco Alberoni — Italy
Igor Bestuzhev-Lada — USSR
Vincenzo Cazzaniga — Italy
Robin Clarke — UNESCO
James Dator — Hawaii, USA
Jacques Delors — France
Pietro Ferraro — Italy
Dennis Gabor — UK
Johan Galtung — Norway
Lars Ingelstam — Sweden
Robert Jungk — Austria
John McHale — USA
Mihailo Markovich — Yugoslavia
Peter Menke-Glückert — FDR
Pietro Pace — Italy
Aurelio Peccei — Italy
Pietro Prini — Italy
Roberto Rossellini — Italy
Valerio Selan — Italy
Jan Strzelecki — Poland
Ota Sulc — Czechoslovakia
Bart van Steenbergen — Netherlands
James Wellesley-Wesley — UK

Scientific Committee

Giorgio Nebbia
(President)
Eleonora Barbieri Masini
(Executive Secretary)
Francesco Alberoni
Jacques Delors
Pietro Ferraro
Dennis Gabor
Robert Jungk
Peter Menke-Glückert
Pietro Pace
Aurelio Peccei
Valerio Selan
James Wellesley-Wesley

IRADES Executive Committee

Eleonora Babieri Masini
(Secretary)
Raimondo Cagiano de Azevedo
M. Letizia Guerrieri
Pietro Pace
Fernando Ragazzini
Carmine Silano
M. Teresa Tavassi La Greca
Carlo Virgilio